Wireless Local Loops

Theory and Applications

Wireless Local Loops

Theory and Applications

Edited by

Peter Stavroulakis
Director, Telecommunication Systems Institute
Crete

JOHN WILEY & SONS, LTD

Chichester • New York • Weinheim • Brisbane • Singapore • Toronto

Other Wiley Editorial Offices

John Wiley & Sons, Inc., 605 Third Avenue,
New York, NY 10158–0012, USA

WILEY-VCH Verlag GmbH
Pappelallee 3, D-69469 Weinheim, Germany

John Wiley & Sons Australia, 33 Park Road, Milton,
Queensland 4064, Australia

John Wiley & Sons (Canada) Ltd, 22 Worcester Road
Rexdale, Ontario, M9W 1L1, Canada

John Wiley & Sons (Asia) Pte Ltd, 2 Clementi Loop #02–01,
Jin Xing Distripark, Singapore 129809

British Library Cataloguing in Publication Data

A catalogue record for this book is available from the British Library

ISBN 0471 49846 7

Typeset in 10/12pt Times by Kolam Information Services Pvt. Ltd, Pondicherry, India
Printed and bound in Great Britain by Antony Rowe Ltd, Chippenham, Wiltshire
This book is printed on acid-free paper responsibly manufactured from sustainable forestry, in which at least two trees are planted for each one used for paper production.

Dedicated to my family
Nina, Peter, Steven, Bill, Stelios
who helped me go when the
going seemed difficult

Contents

List of Contributors

B. R. Badrinath
WINLAB and Computer Science
Rutgers University
73 Brett Road, Piscataway, NJ 08854, USA

Ioannis S. Barbounakis
Telecommunications Systems Institute of Crete (T.S.I.)
Iroon Polytechniou 37 str.
73133 Chania, Crete, Greece
Tel: +30–821–28457
Fax: +30–821–28459
E-mail: ioannis@tsinet.gr

Hans Bhatia
Hughes Network Systems
11717 Exploration Lane
Germantown, MD 20876, USA
E-mail: *bhatia@hns.com*

Roger Easo
Aachen University of Technology
Kopernikusstr. 16
D–52074 Aachen, Germany
Tel: +49 (0)241 88 90 340
Fax: +49 (0)241 88 88 242
E-mail: reo@comnets.rwth-aachen.de

James G. Evans
WINLAB and Computer Science
Rutgers University
73 Brett Road, Piscataway, NJ 08854, USA

Ingo Forkel
Aachen University of Technology
Kopernikusstr. 16
D–52074 Aachen, Germany
Tel: +49 (0)241 88 90 340
Fax: +49 (0)241 88 88 242

Giselle M. Galvan-Tejada
Department of Electronic and Electrical Engineering
University of Bradford
Bradford, West Yorkshire, BD7 1DP, UK

John Gardiner
Department of Electronic and Electrical Engineering
University of Bradford
Bradford, West Yorkshire, BD7 1DP, UK
E-mail: j.d.gardiner@bradford.ac.uk

Dongsoo Har
AirTouch Cellular
2785 Mitchell Drive
MS8–2, Walnut Creek, CA 94598
Tel: +1 510–279–6310
Fax: +1 510–279–6317
E-mail: dhar@nit.airtouch.com

Yoshihiro Hase
Yokosuka Radio Communications Research Centre
Communication Research Laboratory
3–4-Hikarino-oka
Yokosuka-shi, Kanagawa 239–0847, Japan

Masugi Inoue
Yokosuka Radio Communications Research Centre
Communication Research Laboratory
3–4-Hikarino-oka
Yokosuka-shi, Kanagawa 239–0847, Japan

Thomas Jagodits
Hughes Network Systems
11717 Exploration Lane
Germantown, MD 20876, USA
E-mail: *tjagodits@hns.com*

Wha Sook Jeon
Departement of Computer Engineering
Seoul National University
Kwanak-gu, Seoul, 151–742, Korea
E-mail: jeon@ieee.org

Dong Geun Jeong
Departement of Electronics Engineering
Hankuk University of Foreign Studies
Yongin-si, Kyonggi-do, 449–791, Korea
E-mail: dgjeong@ieee.org

Yi-Bing Lin
Department of Computer Science and Information Engineering
National Chiao Tung University
1001 Ta Hsueh Road, Hsinchu 30050, Taiwan
E-mail: *liny@csie.nctu.edu.tw*

Stefan Mangold
Aachen University of Technology
Kopernikusstr. 16
D–52074 Aachen, Germany
Tel: +49 (0)241 88 90 340
Fax: +49 (0)241 88 88 242
E-mail: smd@comnets.rwth-aachen.de

Ian Oppermann
Centre for Wireless Communications (CWC)
University of Oulu
Oulu, FIN-90570, FINLAND
E-mail *ian.oppermann@ee.oulu.fi*

S. Paker
Faculty of Electrical and Electronics Engineering
Istanbul Technical University
Maslak 80626, Istanbul, Turkey
E-mail: spaker@ehb.itu.edu.tr

Peter Stavroulakis
Telecommunications Systems Institute of Crete (T.S.I.)
Iroon Polytechniou 37 str.
73133 Chania, Crete, Greece
Tel: +30–821–28457
Fax: +30–821–28459
E-mail: peter@tsinet.gr

Jaako Talvitie
Centre for Wireless Communications (CWC)
University of Oulu
Oulu, FIN-90570, FINLAND

O. N. Uçan
Faculty of Engineering, Department of Electrical and Electronics Engineering
Istanbul University
Avcilar 34850, Istanbul, Turkey
E-mail: uosman@istanbul.edu.tr

M. Uysal
Wireless Communication Laboratory
Department of Electrical Engineering
Texas A&M University
College Station, 77840, Texas, USA
E-mail: uysal@ee.tamu.edu

Bernhard Walke
Aachen University of Technology
Kopernikusstr. 16
D52074 Aachen, Germany
Tel: +49 (0)241 88 90 340
Fax: +49 (0)241 88 88 242

Gang Wu
Yokosuka Radio Communications Research Centre
Communication Research Laboratory
3–4-Hikarino-oka
Yokosuka-shi, Kanagawa 239–0847, Japan

Howard H. Xia
AirTouch Communications
2999 Oak Rd
MS900, Walnut Creek, CA 94596
Tel: +1 510–210–3461
Fax: +1 510–210–3469
E-mail: howard.xia@airtouch.com

Shun-Ren Yang
Department of Computer Science and Information Engineering
National Chiao Tung University
1001 Ta Hsueh Road, Hsinchu 30050, Taiwan

Preface

This book entitled Wireless Local Loops – Theory and Applications has been motivated by a similar course taught by the editor and by a conference on a similar subject organised by the same editor. The aim of the book is to present in a self-contained manner and in a tutorial format the theoretical aspects and various practical applications of a new medium of transmission which is called Wireless Local Loops which is making a major impact in the provision of new services. Over the years, WLL have been called *radio in the loop* (RITL) or *fixed radio access* (FRA) to indicate a system that connects subscribers to the *public switched telephone network* (PSTN) using radio signals as a substitute for copper for all or part of the connection between the subscriber and the switch. This includes cordless access systems, proprietary fixed radio access, and fixed cellular systems.

Industry analysts predict that the global WLL market will reach millions of subscribers by the year 2000. Much of this growth will occur in emerging economies where half the world's population lacks *plain old telephone service* (POTS). Developing nations like China, India, Brazil, Russia, and Indonesia look to WLL technology as an efficient way to deploy POTS for millions of subscribers – without the expense of burying tons of copper wire.

In developed economies, WLL will help unlock competition in the local loop, enabling new operators to bypass existing wireline networks to deliver POTS and data access. So the question is not will the local loop go wireless, but when and where. It is also expected to play a major role in the third and fourth-generation mobile systems.

As such, it seems appropriate at this point of time to put together a book which will cover the theory and possible applications of WLL which present in a concise and coherent manner concepts which presently can only be found in the proceedings of advanced telecommunication journals and conferences.

Part I covers more of the theoretical aspects which include propagation, modulation, coding, channel modelling, and traffic engineering issues for WLL, where Part II covers prototype designs of various access protocols and a mobility manager as well as a remote management system and provision of multimedia services.

The book starts with a general overview of Wireless Local Loops by I. Barbounakis and P. Stavroulakis, it continues with a presentation of propagation models for WLL by Dongsoo Har and Howard H. Xia. Chapter three covers Capacity Enhancement Issues for Space Division Multiple Access by G. M. Galvan-Tejada and J. G. Gardiner. Modulation aspects are covered in chapter four by O. N. Ucan, M. Uysal and S. Paker. Chapter five covers an algorithmic approach for the generation of orthogonal Hadamard code in order to reduce complexity in low data WLL receivers by J. G. Evans and B. R. Badrinath. Channel modelling for indoors and outdoors WLL is presented in chapter six by Ian Oppermann and Jaako Talvitie. Chapters seven and eight cover the traffic considerations in comparing access techniques for WLL and Dynamic Channel Allocation Schemes by Ingo Forkel, Stefan Mangold, Roger Easo and Bernhard Walke.

Applications are covered in Part II which starts with chapter nine to present an innovative access protocol prototype by Masugi Inoue, Gang Wu and Yoshihiro Hase and chapter ten presents a mobility manager which is based on a private branch exchange. A remote management system is presented in chapter eleven by T. Jagodits and Hans Bhatia and finally the prospects for current and future services using WLL is covered in chapter twelve by Dong Geun Jeong and Wha Sook Jeon. In closing, we believe that this book will help the new generation of graduate students and practising engineers in the coming years to grasp the basic background of WLL which will play a major role in the provision of new services as it relates to third and fourth-generation wireless systems. We also hope that it will assist young researchers to further investigate this new field.

Acknowledgement

I feel indebted to the contributors of this book whose diligent work made this book possible and my staff at T.S.I especially Mr. H. Sandalidis and Chrysanthi Lytra who worked endless hours helping me to put this material together and in a publishing format.

Part I

Theoretical Aspects

1

Introduction to WLL: Digital Service Technologies

Ioannis S. Barbounakis and Peter Stavroulakis

1.1 Background

During the last few years, the telecommunications sector has progressed remarkably thanks to the numerous technological advances occurring in the field. The demand for communication services has also increased explosively worldwide creating or imposing tougher capacity requirements on the telecommunication infrastructure. In developing regions, this demand reflects the great need for the basic telephone services, i.e. the *Plain Old Telephone Service* (POTS), whereas in developed regions it applies to high-rate data and multimedia services at home and/or office. In addition to these recently established conditions, the liberalization of the telecommunication sector, taking place in our days, has unexceptionably driven innovation on the telecommunication infrastructure. It is only the local loop segment that has left unchanged despite all these technological innovations. Lately, however, it has attracted the attention of telecommunication carriers since it proves to be the bottleneck in their network expansion. Consequently, more efficient transmission techniques (ISDN, DSL) improving the capacity of the copper wires or alternative physical media such as fibre, coaxial cable and wireless terrestrial or satellite links have started to be deployed more and more.

It is not only the rapid penetration, which is necessary in developing regions, but also the need for higher capacity in developed regions that have made other physical media apart from our common copper wiring viable solutions in the local loop arena. Today's copper wiring is mostly limited to a maximum distance of 5 km between the subscriber and the local exchange, with the average being in the region of 2 km. This class of transmission channels is sufficient in providing POTS and data through voice-band modems. Moreover, it has reached its upper limits and only thanks to digital techniques such as *Integrated Services Digital Network* (ISDN) and *Digital Subscriber Line* (DSL), it keeps a high competitiveness. ISDN has been the first digital transmission technology to work over existing copper lines offering voice, data and low-resolution video simultaneously. DSL technology has followed offering data and voice integration with a higher efficiency than ISDN but at the cost of farther limitations. DSL lines must be clean copper from the local exchange to the customer premises. The service also degrades dramatically as the distance from the local exchange increases, limiting bandwidth available to customers or preventing access to more rural users. Asymmetric DSL is the technology favoured by many operators or Internet Service and Multimedia Content Providers. Downstream speeds typically are much faster than the upstream speeds. Symmetric DSL is more popular with *local exchange carriers* (LECs), which locally compete with incumbent operators for customers. Connection speeds are the same in both directions.

Optical fibre has been utilized in the trunk network as a more efficient and cost-effective solution for many years now. In many countries it has also replaced copper in the distribution network. However, when considering the local loop the undertaking becomes too risky mainly due to the high cost involved in such a large-scale deployment.

Cable television has become a reality to many people worldwide for more than 20 years now. When the customer base grew up to a significant level, cable operators thought of providing telephony services through a new type of bidirectional cable modem. Although the coaxial cable is a high-bandwidth channel, the fact that only selected areas of the world and selected populations within these areas would be interested in services other than CATV make this medium cost-ineffective for a local loop option.

Another solution, which adopts radio as the transmission medium, in the local loop is the *wireless local loop* (WLL). WLL is often called the *radio local loop* (RLL) or the *fixed wireless access* (FWA). Since WLL is a kind of radio system, it is natural that its technology has been affected by wireless mobile communication technologies. In fact, as will be shown later, most WLL systems have been developed according to the standards (or their variants) for second-generation cellular and cordless systems. However, until now that third-generation cellular systems, i.e. *Universal Mobile Telecommunication Systems* (UMTS) start to be deployed, WLL systems were at a disadvantage compared to their wireline counterparts in terms of voice quality and data rates supported. In general, almost all of cellular/cordless systems or multiple access techniques can be used for narrowband WLL. However, it is also true that there exist some technologies or systems that have comparative advantages in a certain WLL environment.

Many manufacturers and TV broadcasters have been promoting the idea of deploying terrestrial microwave distribution systems mainly for television provision as broadband wireless systems. The philosophy behind such systems is to provide a reverse link as well. Services like Video on Demand and wideband Internet connections are among the first to be offered. At the assigned microwave frequencies, high propagation losses and weather effects such as heavy rain play an important role in the power budget design of the system probably making it a less favoured solution for wireless local loop access in rural and sparsely populated areas.

Last but not least, there are the satellites, which support network access to all subscribers rather than only the fixed ones. Despite the long delays and the high equipment cost, they will play an important role in providing global network access to rural areas not available through other means or small communities with a minimum degree of mobility.

In this chapter, we attempt an overview of several WLL digital service technologies, which have been developed during the last years. We classify them according to their range, capacity and air-interface specifications standardized or not. Through their presentation, our aim is to conclude on what WLL is able to offer to the developing and developed world now and in the foreseeable future.

Section 1.2 outlines the advantages of efficient WLL systems in developing and developed regions. Section 1.3 focuses on the requirements that WLL has to meet in order to compete in the local loop arena. Section 1.4 presents a generic WLL system architecture and focuses on the technological breakthroughs in the wireless transceiver architecture on a per functional block basis. Section 1.5 describes the digital service technologies, which are in the phase of deployment or trial worldwide. Section 1.6 compares all WLL candidate technologies in terms of range, quality, service capability, etc. Finally, concluding remarks are derived in Section 1.7. For completion purposes, two appendices are given. Appendix A clarifies the differences of cellular technologies being deployed with fixed instead of mobile subscribers. Appendix B constitutes an answer to the question 'which multiple access format is more efficient: CDMA or TDMA?'.

1.2 Advantages of Wireless Systems

Wireless systems are justified as a local loop solution because of the cost-effectiveness and/ or limitations of other technologies such as copper, coaxial cable and fibre. However, there has not been established any standard for WLL yet. WLL systems, which are currently deployed, are based on a wide range of radio technologies including satellite, cellular/cordless and many proprietary narrowband or broadband technologies depending on the desired subscriber density as well as on the coverage area under service (see Figure 1.1).

WLL has many advantages from the viewpoints of the service providers and subscribers [1–7]:

Fast Deployment   WLL systems can be deployed in weeks or months as compared to the months or years needed for the deployment of copper wire systems. Faster deployment can mean sooner realization of revenues and reduced time to payback of the deployment investment. Even with higher costs per subscriber that may be associated with the WLL terminal and base station equipment, the faster rate of deployment can permit a higher return of investment. The rapid rate of deployment can also yield first-mover advantage with respect to competitive services, can accelerate the pace of regional economic growth, and can provide substantive progress in the development of needed infrastructure.

Low Construction Cost   The deployment of WLL technology involves considerably less heavy construction than does the laying of copper lines. The lower construction costs may be more than offset by the additional equipment costs associated with WLL technology, but in urban areas, especially, there may be considerable value in avoiding the disruption that the wide-scale deployment of copper lines entails.

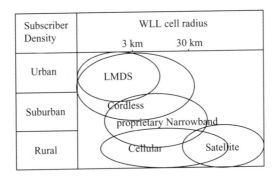

Figure 1.1   WLL coverage using different technologies

Low Operations and Maintenance Cost   The operations and maintenance are easy and the average maintenance time per subscriber per year is shorter 3 to 4 times than their wireline competitors.

Customer Connection Cost   It is low, so overall 'cost per customer' is significantly lower than wireline or cellular systems.

Lower Network Extension costs   Once the WLL infrastructure—the network of base stations and the interface to the telephone network—is in place, each incremental subscriber can be installed at very little cost. WLL systems that are designed to be modular and scaleable can furthermore allow the pace of network deployment to closely match demand, minimizing the costs associated with the underutilized plant. Such systems are flexible enough to meet uncertain levels of penetration and rates of growth.

High Handwidth Services Provision   Using advanced digital radio technologies, WLL can provide a variety of data services and multimedia services as well as voice.

High System Capacity   Among radio systems, WLL enjoys the merits of fixed system: using high-gain directional antennas, the interference decreases. This reduces the frequency re-use distance, increases the possible number of sectors in a sectored cell, and increases, in turn, the system capacity (see Appendix A).

1.3   WLL Service Requirements

The services offered depend strongly on the customer segment. These will in turn impact the bandwidth required to deliver the service and hence the supporting technology, since not all can deliver the high rates required for advanced services.

   The emergence of ADSL, cable network upgrades for data services and developments in 3rd-Generation mobile all impact the WLL service in a competitive environment. They drive the minimum data rate needed for a fixed wireless solution to remain competitive in the residential segment. With the introduction of broadband wireless technologies, data rates of more than 10 Mbit/s are now possible, accommodating bandwidth intensive applications such as video-on-demand or LAN interconnect. The broadband wireless systems being deployed worldwide today are targeting mainly multitenant business buildings with E1/T1 services for aggregated telephony or IP traffic.

   A summary of service needs for different customer types is shown in Table 1.1. In all cases, if WLL systems have to be competitive in service provision to alternative suppliers, they have to satisfy the following requirements that vary with respect to the servicing area, the target group of potential customers and the kind of services offered:

Communications Quality   Since a WLL system serves as an access line for fixed telephone sets, it must provide the same level of quality as conventional telephone systems with respect to such aspects as speech quality, *grade of service* (GOS), connection delay and speech delay.

Table 1.1 Service needs per customer type

Customer\Service	Basic Telephony	Internet data/fax	BRA ISDN	$n \times 64/56$ Kbps	$n \times$ E1/T1 PRA ISDN	LAN ATM	MPEG2	IN functions
Very large business	■	■	■	■	■	■		■
Large business	■	■	■	■	■	■		■
Medium business	■	■	■	■	▼	▼		■
Small Business	■	■	■	■				■
SOHO	■	■	■	■				▼
High spending resident	■	■	■	▼			■	
Med spending resident	■	■	■				▼	
Low spending resident	■	▼						

■ : means full use ▼ : means partial use

Secure Transmission WLL must be secure to give the customer confidence that conversation remains confidential. The system should also include authentication to prevent fraudulent use.

ISDN Support The system should support integrated services digital network (ISDN) when appropriate to provide voice and data service.

Easy Environment Adaptation The system should be capable of small-cell or large-cell operation to serve dense urban or rural areas respectively.

Absence of Interference with Other Wireless Systems A WLL system must not cause any interference with the operation of existing systems, such as microwave communications and broadcasting systems.

High Traffic Volume One characteristic of a WLL system is that it must support a larger traffic volume per subscriber than mobile or even wireline communications systems.

High Capacity and Large Coverage The maximum system range and base station capacity should be large to make the 'cost per subscriber' as low as possible and minimize the entry cost for an operator.

A first assessment of these requirements shows that from the subscriber's perspective service quality and confidentiality as well as bandwidth availability are of great importance. From the perspective of the system operators, the high priority requirements of WLL systems are high-capacity and large coverage. Technically, it is a big challenge to meet these two contradicting sets of requirements and still lower the cost of deploying a WLL system and utilize the spectrum efficiently. Since the three key drivers—voice quality, coverage, and capacity—are always competing among themselves, one may have to determine an acceptable voice quality level first, and then choose a WLL technology that can provide high-capacity and large coverage.

1.3.1   Developing Regions

In many developing regions, the infrastructures for basic telephone services are still insufficient. Accordingly, a lot of population in these areas has not been served with even plain telephony service. For these areas, the requirements of WLL services can be summarized in the following:

- In terms of service coverage, a wide area should be covered within relatively short period.
- Especially, for the regions with dense population, a high-capacity system is indispensable. Here, the capacity is the available number of voice channels for a given bandwidth.
- On the other hand, there may exist wide areas with sparse population. For these service areas, if a small population with low traffic load resides near by, a centralized FSU serving more than one subscriber can be a solution.
- The service fee per subscriber must be low so as to offer the universal service. For this, a high-capacity system is again needed and the cost of system implementation and operation should be low.
- The system should be implemented rapidly so that the services might be launched quickly.

As a trade-off to fulfil the requirements of high-capacity with low service fee, a medium-quality and relatively low data rate of channel (typically, up to 16 kbps) may be unavoidable. Using this channel, only voice and/or voice-band low-rate data communications are possible. However, at the initial choice and installation of WLL system, the service provider should take into account the future evolution of system to provide advanced services.

1.3.2   Developed Regions

In the developed regions, the service requirements contain not only POTS but also other advanced services. It is usual that more than one local exchange carriers and cellular mobile service providers coexist in these service areas. We examine the WLL service requirements from the standpoint of each service provider.

WLL provides a means to establish local loop systems, without laying cables under the ground crowded with streets and buildings. Thus, WLL is regarded as one of the most attractive approaches to the second local exchange carriers. Unfortunately from the second providers' perspective, there are one or more existing providers (i.e. the first providers) who have already installed and operated wireline networks. To meet the increasing and expanding users' service requirements for high-rate data and multimedia services as well as voice, the first providers try to evolve their networks continually (for example, using DSL technologies). The second providers, entering the market in this situation, should offer the services containing competitive ones in terms of service quality, data rate of channel, and supplementary services, etc. That is, the WLL channel of the second provider should be superior to or, at least, comparable with the first operators' one in quality and data rate. Therefore, WLL should provide toll quality voice and at least medium-rate data corresponding to the integrated services digital network (ISDN) basic rate interface (BRI, 2B + D at 144 kbps). In addition, to give subscribers a motivation to migrate to the new provider, the service fee of the second provider needs to be lower than that of the first operators.

Even to the first local switching service providers having wireline networks, WLL can be a useful alternative for their network expansion. Most countries impose the *universal service obligation* (USO) upon the first operators. In this case, WLL can be considered as a supplementary means to wireline networks, for covering areas with sparse population, e.g. islands. The first service requirement for this application of WLL is the compatibility with and the transparency to the existing wireline network. On the other hand, the cellular mobile service providers can offer easily WLL services by using their existing infrastructure for mobile services. In this case, *fixed* WLL service may have competitiveness by combining with the *mobile* services. For example, these two services can be offered as a bundled service [5,8–9]. That is, with a single subscriber unit, a subscriber enjoys the fixed WLL services at home and the mobile services on the street.

1.4   Generic WLL System Architecture

Since WLL systems are fixed, the requirement for interoperability of a subscriber unit with different base stations is less stringent than that for mobile services. As a result, a variety of standards and commercial systems could be deployed. Each standard (or commercial system) has its own air-interface specification, system architecture, network elements, and terminology. Moreover, under the same terminology, the functions of the elements may differ from system to system. In this section, we present a generic WLL architecture (see Figure 1.2).

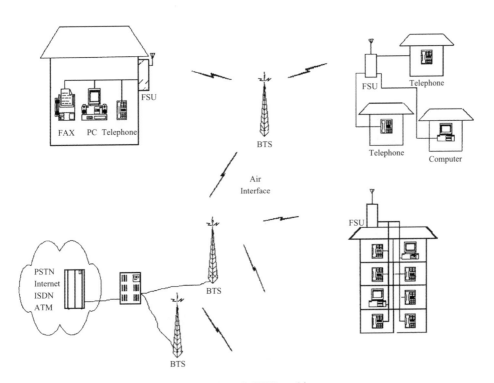

Figure 1.2   Generic WLL architecture

The *fixed subscriber unit* (FSU) is an interface between subscriber's wired devices and WLL network. The wired devices can be computers or facsimiles as well as telephones. Several systems use other acronyms for FSU such as the *radio subscriber unit* (RSU), or the *fixed wireless network interface unit* (FWNIU). FSU performs channel coding/decoding, modulation/demodulation, and transmission/reception of signal via radio, according to the air-interface specification. If necessary, FSU also performs the source coding/decoding. FSU also supports the computerized devices to be connected to the network by using voice-band modems or dedicated data channels.

There are a variety of FSU implementations. In some types of commercial products, the FSU is integrated with the handset. The basic functions of this integrated FSU are very similar to those of the mobile handset, except that it does not have a rich set of functions for mobility management. Another example of FSU implementation is a high-capacity, centralized FSU serving more than one subscriber. Typical application of this type of FSU can be found in business buildings, apartment blocks, and the service area where some premises are located near by (see Figure 1.3).

FSU is connected with the base station via radio of which band is several hundreds of MHz till up to 40 GHz. Since WLL is a fixed service, high-gain directional antennas can be used between FSU and the base station, being arranged by line-of-sight (at least, nearly). Thus, WLL signal channel is a Gaussian noise channel or strong Rician channel (not a Rayleigh fading channel) [6]. This increases drastically the channel efficiency and the capacity of the system.

The base station is implemented usually by two parts, the *base station transceiver system* (BTS) and the *base station controller* (BSC). In many systems, BTS performs channel coding/decoding and modulation/demodulation as well as transmission/reception of signal via radio. BTS is also referred to as the *radio port* (RP) or the *radio transceiver unit* (RTU).

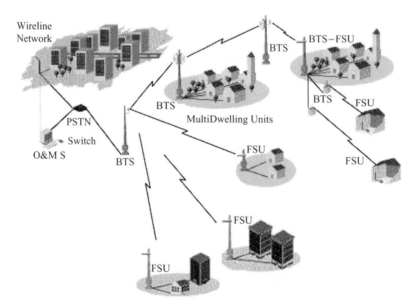

Figure 1.3   FSUs serving multiple subscribers

A BSC controls one or more BTSs and provides an interface to the local exchange (switch) in the central office. An important role of BSC is to transcode between the source codes used in wired network and that at the air-interface. From the above roles, a BSC is often called the *radio port control unit* (RPCU) or the *transcoding and network interface unit* (TNIU).

WLL systems provide fixed wireless access and therefore they do not need to support any mobility features like handover, even though some of these systems are based on cellular standards and products. For a complete comparison between fixed wireless and cellular systems one should refer to Appendix A.

As one can easily understand from Figure 1.2, the WLL services depend not only on the functionality of FSU, BTS, BSC, and air-interface specification but also on the service features provided by the switch in the central office. For example, when WLL is used as a telephony system, there are the basic telephony services (e.g. call origination, call delivery, call clearing, emergency call, etc.) and the supplementary services (e.g. call waiting, call forwarding, conference-calling, calling number identification, etc.). In addition, as in the wired systems, the features such as custom calling features, Centrex features can be supported by the switch [4,6]. If the air-interface provides a transparent channel to the switch, these service features depend totally on the switch functions. So, we hereafter focus on the wireless transceiver functional blocks as well as on the various WLL system technologies rather than the service features provided for by the switches.

1.4.1 Wireless Transceiver Functional Blocks

Thanks mostly due to cellular systems and their penetration worldwide, the wireless transceiver has reached progress levels, which otherwise would be considered intangible. Advances in the areas of antennas, modulations and *digital signal processing* (DSP) have accelerated the design of wireless transceivers into higher levels of functionality, efficiency and signal quality. Below, we distinguish the functional blocks, which a wireless transceiver consists of. We have the chance to present some issues regarding each such functional block.

Antennas Spatial diversity receive antennas are used to combat the flat fading. Directive antennas with a few degrees of beamwidth are generally sufficient to drastically reduce the delay spread, with the drawback of complete outage of transmission if it exists only in a *non-line-of-sight* (NLOS) path. The second drawback is that it can be used only with temporary fixed terminals unless adaptive phase arrays or switchable antennas are used.

Modulators In wireless communications, the decision upon which modulation will be used is very critical. It is not only the capacity that must be offered within the reserved frequency spectrum but also the resistance it has to exhibit to the various types of interference and noise that characterize the wireless channel. The rapid progress in cellular/mobile and *personal communication services* (PCS) has boosted research in this area. First modulations to be adopted in cellular as well as in cordless systems were $\pi/4$-DQPSK (IS-54, Personal Digital Cellular PDC) and *Gaussian Minimum Shift Keying* (GMSK) (*Global System for Mobile communications* GSM). They proved to be the best candidates for the wireless channel where multipath propagation, cochannel interference, fading and shadowing apart from additive noise and intersymbol interference mostly in TDMA dominate. The success of mobile communications has led a whole class of research teams to work upon the standards of the third-generation systems that among the others they will have to support higher data rates. At the same time, new wireless communication systems thrive to find a trade-off between the higher baud-rate and the most efficient data compression keeping the signal quality at acceptable levels. When, however, the baud-rate increases, the multipath effects are intensified. One solution to combat these effects is to increase the symbol length so that it becomes only a fractional time of the mean delay spread. This can be achieved by using either M-ary modulation types (such as B-O-QAM and Q-O-QAM) or multicarrier modulation (such as *offset frequency-division multiplexing*, OFDM), with the drawbacks of increased hardware complexity, increased transmitted power levels, and high-amplitude linear microwave power amplifiers. Alternatively, *direct sequence spread spectrum* (DSSS) resolves the different delays by correlation techniques. However, since digital correlators are generally limited to a few tens of megahertz, and the sequence length has to be greater than 10 to 100 chips, this solution is not always feasible for high data rate transmission even for the case of indoor communications, where the delay spread is on the order of 10–80 ns.

Filters   It is one of the operations cited to be ideally suited for DSP applications. FIR filtering is a repetitive process performed by multiplying the set of input signal samples with a fixed set of known filter coefficients. In the example of IS-136 (TDMA), pulse shaping is done at a transmitter with a square-root raised cosine filter, and appropriate matched filtering has to be done at a receiver. Although straightforward, there are instances when care needs to be exercised to make sure that filtering is executed within the minimum possible number of machine cycles.

Receiver Synchronization Circuits   There exist several layers of synchronization [10]:

— Frequency synchronization
— Carrier recovery for coherent demodulation
— Symbol timing recovery
— Slot and frame synchronization

Most synchronization methods require that they be accomplished through initial acquisition, tracking, and reacquisition. Synchronization of frequency is typically accomplished through the use of *automatic frequency control* (AFC). In digital receivers, the received signal constellation rotation is monitored in a DSP and a phase error based measurement is differentiated and filtered. This error signal is used in a digital VCO to come up with a number, which will correct the operation of an analogue component (VCXO). From the DSP this control signal can be sent through the dedicated D/A to a digitally controlled device or converted to a pulse-width modulation form suitable for transmission as an analogue signal. Algorithms for frequency synchronization are often feedback-based and require the operation of the PLL suitable for DSP implementations. Carrier recovery is associated with coherent receivers, where knowledge of the phase of the received signal is required. It is simple to implement a fully digital phase correction algorithm in DSP firmware, again by monitoring the phase error in a signal constellation. It is the decision of implementers if the phase correction is done in the analogue component based on a digital control signal or fully digitally implemented.

In the example of IS-136, frame timing is the first synchronization that can be accomplished using training symbols at the beginning of all data frames. Frame synchronization is accomplished by correlation of the received waveform with the replica of the training waveform(s) known to the receiver. This is the feed-forward type of operation. The receiver repetitively correlates the signal until it identifies a peak in the correlation function, and based on the peak's location in time adjusts its timing. It is important to note that in cases where a frequency offset exists the correlation will fail when it is done only against the ideal original training waveforms. For good performance in realistic environments it is also necessary to correlate the received signal against frequency-shifted original waveforms. A typical DSP is ideal for correlation operations. Tracking of the frame timing can be accomplished by the same operation, except that the span of the received signal, which needs to be correlated, is significantly smaller since we are close to the actual timing. Here, though, one has the choice of implementing better resolution.

Symbol timing recovery makes sure that a received signal waveform is sampled as close as possible to the optimal sampling point for detection. Since it is desirable to have A/D converters operate at the slowest possible rate, it is required to be able to finely change the sampling position. Indeed, one can choose to adjust the sampling phase of an A/D explicitly. However, more and more often in digital receivers, the preferred choice is to let the converter keep sampling at an arbitrary phase and to use digital interpolation to find the value of the signal at the optimal sampling point from two or more neighbouring samples of an arbitrary phase.

Equalizers The fact that wireless channels have an associated delay spread which causes intersymbol interference requires in some instances that this be compensated for. However, the higher the symbol rate is, the more complex and time- and power-consuming the device is. For a 24.3 kbps IS-136 system, the delay spread which causes trouble and requires an equalizer is on the order of $10\,\mu s$. Two principal techniques are used for equalization: *decision-feedback equalizers* (DFEs) and *maximum-likelihood sequence estimators* (MLSEs). Equalization is one of the most MIPS-intensive functions in cellular phone receivers. Although equalizers are not always needed, since channels often have smaller delay spreads as happens in WLL, receivers/DSPs have to be designed to be able to handle equalization. DFEs consist of two FIR filters and are amenable to DSP implementations. MLSE equalization requires clever memory addressing approaches, which DSPs support.

Channel Coders/Decoders Channel coding is almost always applied in cellular/WLL communications systems. FEC (*forward error correction*) codes and interleaving techniques (to randomize the errors) are used to correct a certain number of bit errors, thus giving a coding gain (relative to the received power). However, this has the drawback of needing an increased transmitted bit rate, leading to higher levels of ISI and creating more decoding delays and complexity due to the necessary interleaving memories. The operation of coding is always simpler than decoding. Both block-code and convolutional-code decoding can be demanding in terms of the number of cycles required. DSP vendors are paying particular attention to efficient software implementations and/or building specialized hardware for trellis search techniques, which are effective for various decoding schemes. These accelerators are probably the first of a number of accelerators that will deal with speeding up the operation of DSPs. It is interesting to note that Viterbi MLSE equalization techniques can sometimes share trellis-searching structures with channel code decoders. This is most obvious in GSM, where modulation is binary.

Automatic Gain Controllers While propagating through a wireless channel, a signal can experience dramatic changes in power levels. Standard deviation of a signal due to shadowing is on the order of 8–12 dB, whereas Rayleigh fading can cause as much as 30–40 dB of rapid signal power fluctuations. It is not always desirable to get rid of all Rayleigh fading fluctuations (especially when they occur rapidly), but shadowing needs to be compensated for. Automatic gain control schemes in modern receivers collect and process data in the digital domain, and then send control information to analogue components, which adjust, signal power levels prior to A/D converters. It is not usual to have overdesigned A/Ds (in terms of the number of bits), which would let DSPs cover most of the dynamic range of radio signals.

1.5   WLL System Technologies

Early WLL systems used standard cellular and cordless technologies to gain access to spectrum. These are at low frequencies, which have become congested and expensive, as mobile operators are able to pay premium rates. In our days, however, WLL deployments also utilize other proprietary systems, narrowband or broadband in frequency bands that have been provided by ITU on a worldwide basis. In general, the frequency bands, which have been used or standardized for WLL service, are described in Table 1.2.

1.5.1   High-Range Cellular Systems

The high-range cellular systems support high mobility and can be characterized by the wider coverage with relatively low data rate. These systems include the second-generation digital cellular systems using 800 MHz band (e.g. IS-95A, and GSM) and their up-banded variations for the personal communications services (PCS) using 1.8–2.0 GHz band (for example, W-CDMA and IS-95B as an up-banded version of IS-95A, and DCS-1800 as an up-banded version of GSM). Since cellular systems are capacity limited due to the limited spectrum resources, they have turned to efficient multiple access techniques such as TDMA and CDMA. Although these two techniques are more efficient than FDMA, it is difficult to say which one is superior. Appendix B touches properly this matter.

Table 1.2   Frequencies used or standardized for WLL

Frequency	Use
400–500 MHz	Rural applications with mostly analogue cellular systems
800–1000 MHz	Digital cellular radio in most countries
1.5 GHz	Typically for satellites and fixed links
1.7–2 GHz	Cordless and cellular bands in most countries
2.5 GHz	Typically for Industrial, Scientific and Medical (ISM) equipment
3.4–3.6 GHz	Standardized for WLL around the world
10 GHz	Newly standardized for WLL in some countries
28 GHz and 40 GHz	For microwave distribution systems around the world

Among the above-mentioned systems, we briefly outline TDMA (IS-136, GSM), and CDMA (IS-95A, IS-95B, W-CDMA) systems [11].

1.5.1.1 TDMA (IS-136/GSM)

TDMA is a narrowband system in which communications per frequency channel are apportioned according to time. For TDMA system, there are two prevalent standards: *North American Telecommunications/Electronics Industry Association* (TIA/EIA) IS-136 and *European Telecommunications Standards Institute* (ETSI) Global System for Mobile Telecommunications (GSM). The IS-136 and GSM standards use different modulation schemes (i.e. $\pi/4$-QPSK for IS-136 and GMSK for GSM). Also, the channel bandwidth of the two systems is different (30 kHz for IS-136 and 200 kHz for GSM). GSM has a frame length of 4.615 ms instead of 40 ms for IS-136. The operational frequencies of these TDMA schemes differ and only GSM supports frequency hopping. GSM uses *Regular Pulse Excitation Long Term Predictive* (RPE-LTP) voice coding algorithm at full rate of 13 kbps or half-rate 6.5 kbps and Enhanced Full Rate at 12.2 kbps whereas IS-136 uses Vector Sum Excited Linear Predictor VSELP at 8 kbps, IS-641-A at 7.4 kbps and US1 at 12.2 kbps. The maximum possible data rate achievable is 115.2–182.4 kbps with General Packet Radio Service supported from GSM and 43.2 kbps for IS-136+. They both use hard handover.

1.5.1.2 CDMA (IS-95A, IS-95B, W-CDMA)

CDMA (IS-95A) is a direct sequence spread spectrum (DSSS) system where the entire bandwidth of the system 1.25 MHz is made available to each user. The bandwidth is many times larger than the bandwidth required transmitting information. In CDMA systems *pseudonoise* (PN) sequences are used for the different user signals with the same transmission bandwidth. For IS-95, a frame length of 20 ms has been adopted. Regarding vocoders, it uses Qualcomm Code Excited Linear Prediction QCELP at 8 kbps, CELP at 8 kbps and 13 Kbps. Compared to the TDMA counterparts, it uses soft handover and either QPSK or O-QPSK as the modulation format.

IS-95-A [12] standard has been developed for a digital cellular system, operating at 800 MHz band. ANSI J-STD-008 [13] being an up-banded variation of IS-95 is a standard for PCS systems, operating at $1.8 \sim 2.0$ GHz band. Recently, IS-95-B [14] merges IS-95-A and ANSI J-STD-008.

IS-95 based CDMA WLL can support two rate sets. A code channel (that is, a traffic channel) operates at maximum of 9.6 kbps with the rate set 1 or 14.4 kbps with rate set 2. Using rate set 1 (rate set 2), the system supports 8 kbps (13 kbps) *Qualcomm's codebook excited linear predictive* (QCELP) vocoder.

IS-95B offers high-rate data services through code aggregation. In IS-95B systems, multiple codes (up to eight codes) may be assigned to a connection. Thus, the maximum data rate is 76.8 Kbps using rate set 1 or 115.2 Kbps, using rate set 2. Since IS-95B can be implemented without changing the physical layer of IS-95A [15], it is relatively easy for the vendor of IS-95 WLL system to develop the IS-95B WLL system.

In mobile IS-95 systems, a sectored cell is designed with three sectors in usual. As mentioned above, in WLL systems, the antennas for BTS and FSU can be arranged by line-of-sight and this reduces interference from the other user. So, the CDMA WLL cell can be designed with six or more sectors [6]. This increases the frequency efficiency and the system capacity.

Both IS-95A and IS-95B have some limitations in supporting high-rate data or multimedia services because of the insufficient maximum data rate per channel. An alternative technology to cope with this problem is the *wideband CDMA* (W-CDMA) [16]. In comparison with the existing narrowband CDMA systems, W-CDMA systems use higher chip rate for direct sequence spread spectrum and, thus, spread its information into wider spectrum bandwidth (typically, equal to or over 5 MHz). Thus, data rate per code channel in W-CDMA can be higher than that in IS-95 systems. Note that all of the major candidates for *radio transmission technology* (RTT) of the *international mobile telecommunications-2000* (IMT-2000) systems have proposed W-CDMA for next-generation mobile communication systems (e.g. [17–19]).

Below, the technical characteristics and the services of a current WLL deployment based on W-CDMA are described. The downlink (from BTS to FSU) uses the band from 2.30 to 2.33 GHz and the uplink (from FSU to BTS) uses the band $2.37 \sim 2.40$ GHz. Thus, the bandwidth of each link is 30 MHz. The spreading bandwidth can be either 5 MHz or 10 MHz. For both spreading bandwidths, the information bit rates are 8, 16, 32, 64, and 80 kbps. For the case of 10 MHz spreading bandwidth, 144 kbps of information bit rate is also available.

The WLL standard defines several options for voice codec: 64 kbps PCM (ITU-T G.711), 32 kbps ADPCM (ITU-T G.726), 16 kbps LD-CELP (ITU-T G.728), and 8 kbps conjugate structure algebraic-code-excited linear prediction (CS-ACELP, ITU-T G.729). However, the service provider seems to offer voice services using 16 kbps LD-CELP and 32 kbps ADPCM since those give toll quality of voice with adequate system capacity. As the voice-band data services, G3 facsimile and 56 kbps modem is planned.

For packet mode data transmission, some dedicated channels, which are separated from voice channels, are provided. They are the packet access channels in uplink and the packet traffic channels in downlink. Using these channels, packet data services up to 128 kbps are offered. In addition, ISDN BRI is also provided.

1.5.2   Low-range Cordless Systems

The advantage of the high-range radio system is the large coverage area of the base stations and the degree of mobility at which access can be supported. The trade-off, however, is low quality voice and limited data service capabilities with high delays. The low-range systems are disadvantaged in coverage area size and user speeds, which is not so important. The advantages include high-quality, low-delay voice and high-rate data capabilities. In comparison with high-range systems, low-range systems provide more wireline-like services. The range of a WLL, however, can be extended via point-to-point microwave hop using a translator which can up-convert signal frequencies in a spectral band to microwave or optical frequencies, and then down-convert to the signal at the remote cell sites before connecting to WLL terminals or buildings. There are several standards for low-range systems. The representative examples are the *digital enhanced cordless telecommunications* (DECT) [11] and its North American variation *Personal Wireless Telecommunication* (PWT), the *Personal Access Communications System* (PACS), and the *Personal Handy-phone Services* (PHS). All these standards adopt the TDMA technology.

1.5.2.1 *Digital Enhanced Cordless Telecommunication (DECT/PWT)*

DECT is a radio interface standard developed in Europe mainly for indoor wireless applications and being deployed for WLL applications as well during the last years. Personal Wireless Telecommunications (PWT) is a DECT-based standard developed by the *Telecommunications Industry Association* (TIA) in the United States for unlicensed personal communications services (PCS) applications. PWT-Enhanced is the version, that is suitable for licensed PCS applications [20].

DECT originally supports small cells (radius of $100 \sim 150$ m) with pedestrian-speed mobility [21]. To use DECT in WLL applications, one of the most important problems to be solved is to extend the maximum coverage of a fixed part (i.e. BTS). A solution is to use directional antennas, by which the maximum diameter of a cell can be extended up to several kilometers. For rural applications, using repeaters at the expense of capacity can extend the coverage [8].

The basic unit of channel in DECT is a time slot per TDMA frame, operating at 32 kbps. If data rates higher than 32 kbps are required, multiple time slots per frame are used. Otherwise, if the requested data rate is lower than 32 kbps, several FSUs can share a 32 kbps channel by skipping time slots. DECT offers toll quality digital speech and voice-band modem transparency either via a 32 kbps ADPCM codec (ITU-T G.726) or as a 64 kbps PCM (ITU-T G.711) bearer service [22]. DECT provides up to 504 Kbps full duplex data transfer and of course BRA ISDN [23]. Since all user information is encrypted, there is confidentiality between the different users belonging to a same cell. DECT has signalling compatibility with basic ISDN and GSM. For more detailed aspects of DECT WLL, one can refer to ANSI J-STD-014 [24].

For Europe, DECT uses Gaussian minimum shift keying (GMSK) with a bandwidth of 1.728 MHz and 12 time slots per carrier. DECT does not efficiently utilize the unlicensed and 10 MHz licensed bands in the United States. Therefore, the protocol was modified to use $\pi/4$ quadrature phase shift keying ($\pi/4$-QPSK) which allows more efficient use of the spectrum.

While other PCS technologies separate the band into a handset transmit band and a base station transmit band (FDD), PWT uses *time-division duplex* (TDD) with both the handset and base station transmitting on the same frequency (at different times). PWT has 24 time slots in 10 ms. Twelve slots are defined for base-to-handset transmission, and 12 are defined for handset-to-base transmission. The overall data rate for voice for handset/base is 32 kbps using *adaptive differential pulse code modulation* (ADPCM), which provides toll-quality voice. The transmission path between handset and base station uses a pair of time slots on the single RF channel.

PWT uses *dynamic channel allocation* (DCA) to assign frequencies to the channels; as the name implies, the frequencies are allocated right before their use. The DCA mechanism provides efficient use of the valuable radio spectrum. The size of the cell covered by an RFP is rather small, less than 150 m for urban applications and 1–2 km for rural applications. For rural applications the coverage can be extended by using repeaters at the expense of capacity. DECT is primarily designed to support pedestrian-speed mobility. This speed is typically less than 10 km/hr.

1.5.2.2 *Personal Access Communication System (PACS)*

PACS employs TDMA/TDM on the radio interface using $\pi/4$-QPSK modulation at a symbol rate of 192 Kbaud [7–8,25–26]. The radio frame is 2.5 ms in duration with

8 bursts/frame. PACS uses *International Telecommunications Union-Telecommunications Standardization Sector* (ITU-T) standard 32 kbps ADPCM speech coder and can maintain very good voice quality with two or three speech coders in tandem. Optionally, 16 Kbps *low-delay code-excited linear prediction* (LD-CELP) being defined as ITU-T G.728 can be used.

For voice-band data, PACS provides 64 Kbps *pulse code modulation* (PCM) connection (ITU-T G.711) by aggregating two time slots. This service is used to support all voice band modems including 56 Kbps modems. PACS supports circuit mode and packet mode data services. In addition, individual message service and interleaved speech/data service are also provided.

- *Circuit-mode data service*: PACS offers reliable real-time data transport service using *link access procedure for radio* (LAPR). LAPR operating in a 32 Kbps channel provides a data throughput of more than 28 Kbps at wireline error rate (10^{-6}).
- *Messaging services*: This is two-way point-to-point message service for large file transfer up to 16 Mbytes. The messages can contain text, image, audio, and video files.
- *Packet-mode data service*: This is a shared, contention-based, RF packet protocol using a data sense multiple access contention mechanism. It supports FSU by using single time slot (32 Kbps) or multiple time slots (up to 256 Kbps) per TDMA frame. The applications being suitable over the PACS packet channel are wireless Internet access and mobile computing, etc.
- *Interleaved Speech/Data*: It provides the ability to transmit both speech information and data information by using a single 32 Kbps time slot. Data is transmitted during the silent period of voice.

Low-power PCS systems, such as PACS, require radio port (RP) operating frequencies to be assigned automatically and autonomously, eliminating the need for manual frequency planning. The automatic frequency assignment in PACS is called *quasi-static autonomous frequency assignment* (QSAFA). QSAFA is a self-regulating means of selecting individual RP frequency channel pairs that function without a centralized frequency coordination between different RPs.

PACS uses *frequency-division duplex* (FDD) for the licensed version and TDD for the unlicensed version. The specification of PACS allows for low-complexity implementations of both *subscriber units* (SU)s and radio ports (RP)s in order to reduce wireless access system costs and network costs. SU peak transmit power is 200 mW, and the average power is 25 mW. The RPs function largely as RF modems, depending on the centrally located RPCUs for most of the functionality traditionally associated with port electronics.

In PACS, the SU determines when and to which RP to perform *automatic link transfer* (ALT) or handoff. The ALT decisions are made by the SU in order to offload this task from the network and to ensure robustness of the radio link by allowing reconnection of calls even when radio channels suddenly become poor. The SU first measures the radio signals. If certain criteria are reached based on the measurements, ALT is performed. The SU determines the new RP for ALT and executes the transfer with the network. FDD provides an advantage over TDD in handling port-to-port synchronization. In TDD operations, both the uplink and downlink of a channel use the same frequency. This requires that RPs be synchronized in order to minimize the interference from each other. In FDD operations, uplink and downlink are on different frequencies. This results in better interference management and does not require adjacent port synchronization, which helps reduce the RF cost of the equipment. The frame delay in PACS is about 5 ms. Such a very low delay negates the need for an echo canceller circuit in the radio equipment. In PWT and other radio technologies, an echo canceller is typically required [8–9].

There are two types of user terminals in PACS/WLL: portable handset (subscriber unit) and fixed access unit. The fixed access units convert the radio signal to a RJ-11 interface signal to the customer premises equipment. The user terminal communicates with the radio port following the JTC/PACS air interface (TDM/TDMA at the 1850–1910 MHz and 1930–1990 MHz frequency bands). The coverage area of a radio port (RP) is 0.5–2 km for the portable handsets and more than 2 km for the fixed access units. The RP connects to the radio port control unit (RPCU) by E1, T1, HDSL, or DSL technologies.

1.5.2.3 Personal Handyphone System (PHS)

PHS is a low-range personal communications services (PCS) technology that was developed in Japan by a consortium of companies to support very-high-density pedestrian traffic and WLL. It is built on a foundation of digital cordless technology and microcell architecture.

PHS uses $\pi/4$-DQPSK in the RF band of 1900 MHz as DECT with a bandwidth of 300 KHz per channel. Each channel consists of either 3 or 4 time-slots. However, it has a better spectrum efficiency than DECT since it has 4 time slots per 300 KHz carrier instead of 12 slots per 1.728 MHz carrier. The multiple access scheme used is TDMA/TDD and the voice coding is 32 kbps ADPCM. PHS makes use of Dynamic Channel Assignment and is more flexible in network planning and more cost-effective and suitable for WLL. Due to its architecture, it is less sensitive to multipath and delay and has bigger cell coverage. In urban areas, it uses 5 times fewer base transmission stations than DECT when covering the same area, whereas in rural areas PHS has a wider service coverage than DECT since it is more tolerant of delay spread.

PHS *personal stations* (PSs) consist of handheld units that can operate as simple cordless phones, as transceivers for communications with other personal stations, or as mobile terminals to access the *public switched telephone network* (PSTN). The mode of operation must be selected by the user. The *cell stations* (CSs) handle the control and transmitter functions. The CS consists of the antenna and the base station unit. Its output power ranges from 100 mW to 500 mW according to the number of users in the area to be served. CSs are usually mounted on utility poles, payphone boxes, and roofs. The CS is connected to the fixed network with integrated services digital network (ISDN). The control station is essentially a database unit for storing subscriber data.

1.5.3 Proprietary Narrowband WLL

The number of competitors in the local loop and service capability, influences likely penetration, and hence capacity and range requirements of the technology solution. Among the several WLL systems already in use in various markets, we outline the TDMA proprietary systems E-TDMA of HNS and Proximity I/II of Nortel, the CDMA proprietary systems QCTel of Qualcomm, Airloop of Lucent and Airspan of DSC, and finally the FH-CDMA/TDMA system Multigain of Tadiran.

1.5.3.1   HNS E-TDMA

E-TDMA [27] is an extension to the IS-136 cellular TDMA standard that provides support for WLL with increased capacity and improved network performance while maintaining the large coverage area feature of other cellular standards. E-TDMA offers a choice of subscriber unit platforms including *single subscriber units* (SSU) and *multiple subscriber units* (MSU) capable of supporting up to 96 lines, depending on the subscriber traffic load and MSU provisioning. The single subscriber units support high-capacity digital voice, fax, and data transparently using a standard RJ-11 interface, and enable multiple terminal connections as simple extensions on a single access unit or per directory number. Such units are appropriate for locations with low population densities such as residences and small businesses. Multiple subscriber units provide access to the WLL system in areas of high population densities such as hotels and apartment buildings. MSU and radio resources are allocated on a call-by-call basis, thereby reducing the required hardware.

The E-TDMA base station provides an improved control channel to dynamically assign channels and time-slots to active speakers. A 5 kbps rate voice coder is also used which more than doubles the capacity over IS-136. Finally, the implementation of *discontinuous transmission* (DTX) along with *digital speech interpolation* (DSI) means that both the base station and the subscriber station transmit only when speech is present (about 40 percent of the time), thus sharing the radio resource effectively with other users. E-TDMA supports a wide variety of country variant signalling. Tones and line signalling variations are software programmable, and in a number of cases can be set via system parameters. Both 16 kHz metering and Polarity reversal signalling mechanisms for pulse signalling can be supported, if they are generated and supported by the switching system. Thus, E-TDMA can interface with a wide variety of metering and public pay phone equipment. Depending on the subtending switching equipment, E-TDMA is capable of supporting virtually all of the vertical features and CLASS features recommended by TR-45, as listed above, including call waiting, call forwarding, conference calling, and so on. The main strengths of cellular-based WLL systems over low-range PCS based WLL systems include coverage, speed of deployment, and spectrum efficiency. The fundamental disadvantage is the limited range of available user bandwidth. This trade-off implies a market for both system types.

1.5.3.2   Nortel Proximity I/Proximity II

Nortel makes a class of narrowband WLL systems under the banner Proximity which are not based on cordless or cellular technologies. Proximity I is a proprietary TDMA system developed in conjunction with the United Kingdom WLL operator Ionica, one of the first operators in the world to deploy a proprietary WLL system. Proximity I offers a wide range of services, including 64-kbps voice and data links and a second-line capability. Subscriber units link to base stations over the air-interface, and base stations then are connected directly back to a PSTN switch.

Proximity II is an upgraded version, that is more flexibly tailored to suit each individual operators' requirements—from small city based systems supporting a few thousand customers up to large nationwide systems with capacities in excess of one million. Proximity II provides also BRA ISDN service and enables high-rate Internet access at 128 Kbps. Its compact base station has a capacity of 2000 lines and it may be located up to 40 km from the users. The user premises equipment supports one or two lines for PSTN or ISDN terminals. Its System Management is compatible with public network switches through V5.2 signalling.

Both system versions use TDMA channels of bandwidth 3072 kHz in a cluster size of 3, and *quadrature phase shift keying* (QPSK) modulation format. Up to 54 TDMA bearers can be accommodated in the 3.4–3.6 GHz assignment using frequency division duplex (FDD) either 50 or 100 MHz with a maximum of 18 channels on any given base station. DCA is not provided, but it is relatively easy to reconfigure the frequency assignment from the operations and maintenance centre.

1.5.3.3 Qualcomm QCTel

Qualcomm's QCTel CDMA WLL System is a Fixed Wireless Access WLL [28]. A basic six-sector QCTel system may support 24 000 subscribers. The QCTel technology supports 8 kbps voice and up to 7.2 kbps data rate. QCTel supports limited mobility, and the subscriber unit can be a portable handset. The handset communicates with the base station transceiver using the IS-95 air-interface (CDMA/FDD at the 800 MHz, 900 MHz, and 1.8–2.2 GHz frequency bands). The handset can support multiple lines. The transmit power is 2 W (with power control).

The base transceiver station (BTS) communicates with the handset using the IS-95 air-interface. The maximum transmit power is 50 W. The cell range is 25 km. The capacity is up to 45 voice channels. Up to 20 BTSs may be collocated with the base station controller (BSC) at the central office. Or 30 BTSs per area may be connected to a BSC using the T1/E1 technology (up to three areas). The BSC is collocated with a central office, which connects to a switch of the PSTN using T1, E1, T3, or E3 digital multiplexed trunks. The call control is done by R2 inband signalling, and the OMC signalling is done by SS7 or X.25.

1.5.3.4 Lucent Airloop

The Lucent Airloop technology is another proprietary CDMA-based system developed for a wide rank of customers. It operates mainly in the 3.4-GHz band using 5-MHz wide channels, each supporting 115 16-kbps channels. To support 32-kbps ADPCM, two channels are used simultaneously. The spreading code is 4096 kbps; thus, for a 16-kbps data rate, a spreading factor of 256 is used. The system employs a network of *radio base stations* (RBSs) to provide coverage of the intended service area. The main functional blocks of the network are the following:

- *Central Office* (CO): It contains digital switching and network routing facilities required connecting the radio network to ISDN and the Internet.
- *Central Access and Transcoding Unit* (CATU): It controls the allocation of radio resources and ensures that the allocation is appropriate to the service being provided, for example, 64-kbps digital, 32-kbps speech, ISDN. It also provides transcoding between various speech-coding rates and the switched 64-kbps PCM.
- *Central Transceiver Unit* (CTU): It provides the CDMA air-interface. It transfers ISDN and *plain old telephony* (POTS) signalling information transparently between the air-interface and the CATU.
- *Network Interface Unit* (NIU): It connects the subscribers to the radio network through two functional blocks, the *intelligent telephone socket* (ITS) and the *subscriber transceiver unit* (STRU).

- The ITS provides the point of connection to the subscriber's terminal equipment, for example, PABX, telephone, or LAN.
- The STRU is located on the outside of the subscriber's building and consists of an integrated antenna and radio transceiver. The STRU provides the interface between the ITS and the CDMA air-interface. The STRU is connected to the ITS by a standard four-wire telephone or data networking cable.

The type of service being provided by the connection determines the number of subscriber connections supported by each NIU. The basic NIU connection provides a single ISDN (2B + D) connection, effectively giving two unrestricted 64-kbps channels. The same unit also can he configured as either two or eight individual POTS lines using ADPCM and *Code excited linear predictive* (CELP) speech coding, respectively.

The modulation technique employed first takes each 16-kbps channel and adds error-correction coding to reach 32 kbps. It then uses Walsh spreading with a spreading factor of 128 to reach the transmitted data rate of 4 Mbps. Finally, it multiplies that by one of a set of 16 PN code sequences also at 4 Mbps, which does not change the output data rate but provides for interference reduction from adjacent cells. During design of the network, each cell must have a PN code sequence number assigned to it such that neighbouring cells do not have the same number.

1.5.3.5 DSC Airspan

The DSC Airspan system was developed in conjunction with BT, which is using the system for rural access at 2 GHz. The system provides 64-kbps voice channels and support for up to 144 kbps ISDN services. DSC claims a cluster size of between 1 and 3, depending on the environment. Voice currently is provided using 64-kbps PCM and ADPCM at 32 kbps. The system currently provides 2B + D ISDN per subscriber or, alternatively, 2×64 kbps data channels. Up to 6×64 kbps data channels can be achieved by combining three subscriber units.

Radio channels are 3.5 MHz wide. Each 3.5-MHz channel provides up to fifteen 160-kbps radio bearers. With the current deployment, each 160-kbps bearer can provide two 64-kbps voice-channels or four 32-kbps voice channels, each to a different house.

1.5.3.6 Tadiran Multigain

Tadiran markets its proprietary system as FH-CDMA/TDMA. In the Tadiran system, users transmit in a given TDMA slot. However, the actual frequency in which they transmit changes from burst to burst, where a burst lasts 2 ms (hence, there are 500 hops/s). In a given cell, no two users transmit on the same frequency at the same time. However, users may transmit on the same frequency in adjacent cells. By employing different hopping sequences in adjacent cells, if a collision does occur it will be only for a single burst. Error correction and interleaving largely can overcome the effect of such a collision. The system has the advantages of the simplicity of a TDMA system coupled with some of the 'interference-sharing' properties that make CDMA spectrally efficient. Tadiran claims that a cluster size of 1.25 can be achieved by that approach when directional antennas are deployed. In practice that seems somewhat optimistic, and it might be expected that a cluster size of 2 would be more realistic.

The system uses a voice coder of 32 kbps. It employs TDD, in which both the uplink and the downlink are transmitted on the same frequency but at different times. Each 1×1 MHz channel supports eight voice channels. Hence, 16 voice channels per 2×1 MHz can be supported before the cluster effect is taken into account, and assuming a cluster size of 2, around 8 voice channels/cell/2×1 MHz.

1.5.4 Proprietary Broadband WLL

A number of systems have been proposed and implemented that use WLL techniques to deliver broadcast TV [29]. Such systems fall into a niche somewhere between terrestrial TV broadcasting and WLL telephony delivery. They offer advantages over terrestrial TV broadcasting in that they can provide many more channels and may offer advantages over the other WLL systems discussed here in their ability to deliver high-bandwidth services. Broadly speaking, the only difference between WLL and microwave distribution is that the latter tends to be transmitted at much higher frequencies, such as 40 GHz, where significantly more bandwidth is available and hence wider bandwidth services can be offered. However, such higher frequencies result in lower propagation distances and more costly equipment. Further, rain fading can be a significant problem in some regions at such frequencies, making reception unreliable.

The first microwave distribution systems were implemented in the United States, where they were called *microwave multipoint distribution systems* (MMDS). The systems tended to operate between 2.15 GHz and 2.682 GHz and provided a maximum of 33 analogue TV channels to communities at a bandwidth of 500 MHz. They were entirely broadcast systems, and no return path capability was provided. However, compared to cable TV and satellite systems that support 30–60 and 150–200 video channels respectively, MMDS operators had to resort to digital compression techniques to become competitive. Many such systems are still in existence between 2.5 and 3 GHz, but with the introduction of digital broadcasting they will become increasingly outdated.

After MMDS, digital distribution systems operating at around 29 GHz were introduced in the United States and the Asia–Pacific countries. The systems, known as *Local multipoint distribution systems* (LMDS) [30], can provide many more channels with a higher quality but lower range. It is considered as a strong candidate for next-generation *broadband WLL* (B-WLL) services. The spectrum for LMDS differs from country to country but it is usually in the $20 \sim 30$ GHz band. The LMDS applications include a variety of multimedia services such as POTS, ISDN, *broadband ISDN* (B-ISDN), television program distribution, videoconference, VOD, teleshopping, and Internet access. LMDS can offer two-way wireless services, whereas MMDS and satellite systems require terrestrial wired networks to communicate back to the headend, for example, to select programming or use VCR-type controls on *video-on-demand* (VOD) programming.

The primary disadvantages of both the MMDS and LMDS are cochannel interference from other cells and limitations on coverage (up to 25 miles for MMDS and up to 5 miles for LMDS) [31]. Millimetre-wave radio signals do not penetrate trees. Thus, line-of-sight propagation paths are required. This requirement can make antenna placement on subscriber homes challenging. Despite, however, the fixed locations of both transmitter and receiver, the influence of motion of traffic and foliage, even in a line-of-sight location, creates a fading environment, which is much more hostile than measured for conventional cellular mobile systems, at say, 2 GHz. Temporal fades of over 40 dB at the rate of at most a few (<2) hertz are frequently seen in these environments, imposing very stringent requirements on the error correction coding of the transmitted bitstream.

Another serious problem in wireless multimedia services is traffic asymmetry between uplink and downlink [16–18,32]. For example, let us consider Internet access or remote computing. In these applications, short commands are transmitted via uplink, whereas relatively large files are transmitted via downlink. In these cases, if both links have the same bandwidth, the system capacity can be limited by the downlink. This, in turn, results in bandwidth waste of uplink and, eventually, spectrum inefficiency. To cope with traffic unbalance, the spectrum allocation for LMDS is given to be asymmetric between uplink and downlink. Since LMDS is a FDD system, the downlink bandwidth should be appropriately wider than the uplink. Another solution for this is to use time division duplex (TDD) between two links. Thus, CDMA/TDD systems, having both the merits of CDMA (e.g. in capacity) and the advantages of TDD (e.g. flexibility in resource allocation), have been attracting many researchers' attention recently [16–18].

Similar initiatives in the United States and Europe have lead to the system operating at 40 GHz, *Microwave Video Distribution System* (MVDS). Early in the development process of MVDS, it became apparent that to provide competition to cable and to maximize the revenues that could be achieved, a return path from the home to the network would be required. That allows voice and limited data enabling, for example, selection of video films. Systems capable of providing such a return path are now in a trial stage. The return path can provide around 20 kbps of data.

MVDS systems still are relatively immature, so it is difficult to provide significant amounts of information on particular products. It can be seen that in terms of telephony provision, MVDS is inferior to other WLL systems. However, when telephony is viewed as a service offered on the back of video distribution, it looks more attractive. It is too early to say whether the economics of MVDS will allow the telephony component to be sufficiently cheap so that users will accept its relative shortcomings. However, for many users, telephony is a critical service that they will not compromise to realize some savings. Hence, at least for the next few years, it is unlikely that microwave distribution systems will provide an acceptable WLL service. Instead, they provide a wireless alternative for the cable operator.

1.5.4.1 HNS AIReach Broadband

AIReach Broadband constitutes a powerful platform for offering fibre-quality 'last-mile' wireless solutions that encompass voice, video, data, multimedia, and Internet services. It aims at serving either individual business customers or multitenant offices/residential complexes. The AIReach Broadband product family consists of two product series. One is ideally suited for semi-urban or light to medium urban areas whereas the other is suited for medium to high-density urban areas. They both address the small to medium size business customers and *multidwelling unit*s (MDUs) and they operate either in the ITU/ETSI frequency bands: 3.5 GHz, 10.5 GHz, and 24–26 GHz or in the frequency bands between 24–42 GHz. The maximum data rate achieved per carrier is 4 Mbps and 45 Mbps respectively.

AIReach Broadband system can start with a single radio at the hub and as few as two subscriber terminals though a completely scaleable hub. It uses 64-QAM as modulation format achieving one of the highest spectral efficiencies available. Moreover, the high-capacity product series deploys *Tri-Mode Modulation* (TMM), which enables balance between coverage and capacity, since with TMM, a single hub radio can switch modulation modes on a burst-to-burst basis.

AIReach Broadband assigns bandwidth on demand as well as voice and data concentration via dynamic bandwidth management making the economics very competitive to wireline solutions. Outdoor terminals take up less rooftop space and provide better installation options. Small indoor terminals with easy front access for all of the cabling are designed for tight spaces and cluttered environments in telecom closets. The low-capacity product series offers the following services and interfaces: E1; N × DS0; Ethernet and Frame Relay; services with BoD; V5.2; Internet Services; BRA ISDN; ATM; 10 BaseT; V35/X.21. Additionally, the high-capacity product series offers: N × T1/E1; LAN; ATM; 10/100 BaseT; PRA ISDN; MPEG2.

1.5.4.2 Motorola SpectraPoint

SpectraPoint has a strong headstart in the direction of facilitating multiplatform integration via IP and ATM, having been working with Cisco Inc. for most of the last few years to integrate router switches and other IP components into the LMDS access system. SpectraPoint already integrates its product series with software, which supports dynamic changes in modulation from the more robust, low bits-per-Hertz levels to the noise-sensitive, high-capacity levels as weather conditions change or customers shift the trade-offs they want to make between bandwidth efficiency and quality of service.

The air-interface of the SpectraPoint platforms already supports QPSK (quadrature phase shift key) and 8, 16 and 32 QAM (*quadrature amplitude modulation*) modulation formats as well as Viterbi and Reed Solomon as Forward Error Correction methods. Frequency re-use is enhanced via polarization diversity (factory-selectable horizontal or vertical). The channel spacing is 40 MHz allowing for a 45 Mbps downstream speed and a 2 up to 10 Mbps upstream speed. The average transmitted power is 1 W for the base station and 100 mW for the subscriber unit.

One of the innovations Spectrapoint brings to LMDS fixed wireless products is the ability to transport everything in ATM while using time division multiple access to dynamically alter the amount of bandwidth devoted to any one user's needs. This way, all the users on a single RF LMDS channel, which now supports up to 45 Mbps and in due course 155 Mbps, can pay for services on an as-needed basis, allowing service providers to more efficiently allocate bandwidth, that is not in use from one moment to the next.

1.5.4.3 Nortel Reunion

Reunion is another *point-to-multipoint* (PMP) system—also referred to as *broadband wireless access* (BWA). It is similar in design to cellular or narrowband wireless local loop systems, but offers bandwidth connection ranges from 64 Kbps up to 155 Mbps—offering great flexibility in serving local access markets. Reunion's unique Quad-4 architecture exploits the potential of four access technologies to produce exceptional network flexibility and efficiency as well as to provide consistency and congruence with wired networks. The advantage of Quad-4 is that it is able to tailor and optimize the network deployment. Quad-4 makes FDMA, TDMA, ATM, and IP connections possible—all from a single platform.

- FDMA provides efficiency in delivering high volumes of data.
- TDMA's greater spectrum and cost efficiency is suited for low bandwidth and sporadic voice and data needs.

- ATM is an excellent solution for multimedia applications and high Quality of Service requirements.
- IP is the technology of choice for low-cost customer premise equipment and end-user applications to help penetrate the SOHO market.

Reunion portfolio offers High-Rate Data Transfer, LAN/WAN Interconnection, Internet/Intranet Access, Telephony, Voice over IP, Corporate Video Services, Home Banking, Distance Learning, Tele-Medicine, Video Conferencing, VPN, E-Commerce, Web TV, Interactive Gaming, Video Streaming. Reunion can be deployed to handle bundled multimedia services or single service solutions.

The Reunion network architecture consists of the following three major elements:

- The Reunion Base Station, consisting of the *Network Node Equipment* (NNE) and the *Base Station Transceiver* (BTR), facilitates the multiplexing, mapping, modulation, and transmission of multimedia content to and from the access market. This equipment, which operates in a variety of downstream and upstream frequencies between 2 GHz and 42 GHz, offers high-capacity. All of this capacity and performance is packed into a small footprint.
- The Reunion element management system facilitates the operations, administration, maintenance, and provisioning of the network.
- A cost-effective integrated CPE meets the needs of the small- to medium-sized customer, providing up to four E1/T1 and 10 Base-T circuits, which utilize either TDMA or FDMA access technologies. A modular CPE, expandable to accommodate future needs, is used to serve building sites with multiple small to medium tenants as well as larger, more bandwidth-intensive customers.

1.5.4.4 Alcatel Evolium

The Alcatel LMDS Solutions provides broadband last mile connections to thousands of subscribers from a single hub. It provides a wireless local infrastructure, using line-of-sight radio links over distances of up to 5 km, handling full two-way communications for more than 4000 end users, delivering true broadband capacity, at a bit rate of up to 8 Mbps through a wide variety of narrow- and broad-band communications services.

Alcatel LMDS utilizes co-polarized or cross-polarized, single carrier or multicarrier radio solutions to get the most out of the allocated spectrum. It supports a Guaranteed Cell Rate (capacity available at all times) and a Peak Cell Rate (maximum capacity available whenever there is spare capacity) on a per customer basis.

The air-interface makes use of a patented TDMA frame, which is optimized for any mix of circuit and packet data applications, with real time Dynamic Bandwidth Allocation. Among its advantages, there exist fibre-like reliability, 99.995 % availability, 10^{-14} BER, encryption over the air and on-line system upgrades with new features by in-band software download.

Alcatel LMDS operates in frequency bands between 10 and 43 GHz. Services include:

- circuit-switched voice, data or mixed voice/data;
- distributed or centralized architecture;
- multiservice cross-pole or mono carrier pole;
- virtual leased line (T1/E1 or N × 64 Kbps);
- IP/Ethernet/ATM/Frame relay;
- bandwidth on demand.

The three components—Base Station, Customer Terminal Station and Network Management—constitute a star-architecture network which can be configured and reconfigured to meet the current and future access network requirements.

Base Station Each base station consists of a Radio Base Station (RBS) and a *Digital Base Station* (DBS) and serves as a hub for as many as 4,000 Network Terminations, transparently handling a virtually infinite variety of voice and high-rate data services. The base station is connected to the switching and routing platforms via any standard high-capacity transmission link.

Terminal Station Each Customer Terminal Station consists of a small (26 cm diameter) outdoor solid-state antenna (Radio Termination) and a simple interface unit (Network Termination). The terminal station is connected to a base station by a line-of-sight digital radio link.

Network and Service Management The Network Service Manager is a highly integrated open architecture solution for managing multitechnology, multiservice networks on a single platform. It extends the management reach of the operator from the wireline network in an integrated fashion into the broadband wireless access network.

1.5.5 Satellite-based Systems

These systems provide telephony services for rural communities and isolated areas such as islands. Satellite systems are designed for a Gaussian or Rician channel with K factor greater than 7 dB. These systems can be either of technology designed specifically for WLL applications or of technology piggybacked onto mobile satellite systems as an adjunct service [5]:

Of these, the former offers quality and grade of service comparable to wireline access, but it may be expensive. The latter promises to be less costly but, due to bandwidth restrictions, may not offer the quality and grade of service comparable to plain old telephone service (POTS). An example of a satellite-based technology specifically designed for WLL is the *Hughes Network Systems* (HNS) *telephony earth station* (TES) technology. This technology can make use of virtually any *geostationary earth orbit* (GEO) C-band or Ku-band satellite. Satellite technology has been used to provide telephony to remote areas of the world for many years. Such systems provide an alternative to terrestrial telephony systems where landlines are not cost-effective or where an emergency backup is required. There are several proposed systems for mobile satellite service, including the *Inmarsat International Circular Orbit* (ICO) system, Globalstar, and *American Mobile Satellite Corporation* (AMSC) system. These systems are specialized to support low-cost mobile terminals primarily for low bit rate voice and data applications. Fixed applications are a possible secondary use to mobile applications. There is a great deal of difference between these systems, especially when considering the orbit and the resultant propagation delay. The number of satellites and the propagation delay pose very different constraints on system design, so that there is no true representative system. For example, GEO satellite systems are not required to support handover even for most mobile applications. *Mid-earth orbit* (MEO) and *low earth orbit* (LEO) satellite systems require handover capability for all fixed and mobile applications because the satellites are in motion relative to the earth's surface even when the terrestrial terminal is fixed. This can be problematic if the handover is supported in the switch because *mobile switching centres* (MSCs) support sophisticated mobility functions such as link handover, but do not typically support ordinary switching functions such as hunt groups, for example, which are highly desirable in a WLL system.

1.5.5.1 HNS Terminal Earth Station Quantum System

The HNS terminal earth station (TES) product is used for the Intelsat network to provide remote-access telephone service [33]. It is one of those systems utilizing geostationary satellites and can offer wireless local access in areas where copper cable is difficult or even much expensive to use. The TES system is a satellite-based telephony and data communications network providing mesh connectivity between multiple earth stations. The system provides call-by-call *demand-assigned multiple access* (DAMA) circuits and pre-assigned circuits, via single hop *single channel per carrier* (SCPC) communications paths between earth stations. It supports both public and private networks, and is capable of operating with any telephony interface from individual subscribers to toll switches and major gateways.

An outdoor RF terminal and antenna plus indoor IF and baseband equipment perform the wireless access subscriber unit functions. *High-power amplifier* (HPA) power options include 5, 10, and 20 W for C-band and 2, 5, 8, and 16 W for Ku-band. A small reflector antenna of 1.8–3.8 m diameters is required. TES remote terminals communicate with each other and the *network control system* (NCS) using virtually any Ku- or C-band satellite, using single channel per carrier access to the satellite.

The air-interface employs quadrature phase shift keying (QPSK) or *binary phase shift keying* (BPSK) modulation, depending upon the user information and coding rates. FEC is provided at rate 1/2 or rate 3/4. Scrambling is used to spread the transmitted energy across the satellite channel bandwidth. Differential coding resolves phase ambiguity in the demodulated signals. Voice is coded using 32 kbps adaptive differential pulse code modulation (ADPCM) (ITU-T G.721) or 16 kbps low delay-code excited linear predictive (LD-CELP) (ITU-T G.728).

The terrestrial interfaces toward the user which are supported by TES include four- and two-wire *ear and mouth* (EM) or *single-frequency* (SF) inband signalling, RS-232, RS-449, and V.35 data interfaces. In addition, single-line versions supporting two-wire, RS-232, and ISDN interfaces are provided whereas multiline access to a PBX can also be supported. The TES ISDN earth station provides 56 kbps pulse coded modulated (PCM) voice. The wireless access network unit equipment logically includes the satellite, terminal equipment, and the Network Control System. Voice calls and asynchronous data calls can be made on demand under the control of the centralized DAMA processing equipment of the NCS. Satellite channels for user information are allocated only for the duration of these connections.

TES supports telephony, synchronous and asynchronous data, facsimile, ISDN BRI data, and E1 and T1 trunking between remote terminals anywhere in the system. Voice and data traffic is transferred directly between remote terminals, not via the NCS, to minimize the delay using a single satellite hop. Features and services are based on the remote PBX rather than on the centralized PSTN interface, which is an E1 or T1 trunk.

1.6 Comparison of WLL Services

Wireless local loop as an access network technology competing with its wire-line counterparts, has to support the full range of Telecommunication Services: Telephony, Fax, Voice-band Data Modems, Leased Line Data and Basic Rate ISDN. It has to be fully compatible with the public network and fully transparent with other access equipment. It must ensure an inherently secure radio interface and simple Installation, Operation and Maintenance procedures. Apart from circuit-switching services, WLL must be capable of offering packet-switching data services such as X.25, X.21, Frame Relay and IP as well. Particularly, the broadband systems must support broadband standards like ATM and SDH to connect transparently to the installed wireline network infrastructure, Figure 1.4.

Table 1.3 gives a summary of the WLL technologies mentioned in the previous sections. As shown in the table, all systems with the exceptions of 900 MHz cellular ones offer toll quality voice services and medium to high-rate data services. Among these systems, cordless and proprietary broadband wireless access, are characterized by short-range radio technology and very high-rate data services. Especially, the broadband wireless access systems are of equal capacity to their broadband wireline counterparts, fibre, coaxial cable and ADSL. These characteristics make such systems as the most appropriate WLL choice for developed regions and/or especially urban/suburban areas.

On the other hand, cellular and satellite WLL systems are characterized by high-range radio technology but low to medium-quality voice and low to medium-rate data services.

Table 1.3 Summary of WLL services

	Cordless	Cellular		Proprietary Narrowband (FWA)	Proprietary Broadband (BWA)	Satellite-Based Systems
		2nd Generation	3rd Generation			
	DECT/ PACS/ PHS	GSM/IS-95	W-CDMA	Proximity/ Airspan/ Multigain	Allreach/ Spectrapoint/ Reunion/ Evolium	HNS Quantum System
Voice Codec	32 Kbps ADPCM 64 Kbps PCM	13 Kbps RPE-LTP/ 13 Kbps CELP	64 Kbps PCM 32 Kbps ADPCM 16 Kbps LD-CELP	64 Kbps PCM 32 Kbps ADPCM	32 Kbps ADPCM	32 Kbps ADPCM 64 Kbps PCM
Data rate up to/ Service capability	504 Kbps/ 256 Kbps BRA ISDN	9.6 Kbps 182.4 Kbps (GPRS)*/ 115.2 Kbps (IS-95B)*	114 Kbps BRA ISDN	128 Kbps BRA ISDN	45 Mbps (Downlink)/ 10 Mbps (Uplink) PRA ISDN	144 Kbps BRA ISDN
Range/ cell size	Low/small	High/large	High/large	Medium	Low/small	Very High/very large
Quality	Excellent	Medium	Medium	Excellent	Excellent	Medium
Frequencies	Standard	Standard	Standard	Standard	Standard	Custom

* Only with channel aggregation techniques.

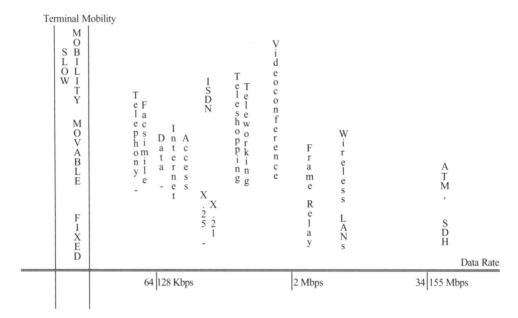

Figure 1.4   WLL applications supported at various terminal speeds and data rates

Hence, they are the appropriate choice for developing regions and/or rural areas and villages. Last, medium-rate data services and toll-quality voice characterize proprietary narrowband WLL systems. Their range can be as high as 40 km, which makes them ideal for deployments without capacity limitations. Finally, only the proprietary broadband systems, known also as LMDS or Point-to-Multipoint (PMP), offer local loop access for future multimedia applications such as video distribution.

Appendix A   Fixed versus Mobile Cellular Systems

WLL is only a fixed wireless communication system using implemented wireless technologies to compete its wireline counterparts. It has the philosophy of fixed communication systems regarding terminal mobility at both ends of the transmission link such as terrestrial microwave point-to-multipoint and satellite microwave systems. Hence, due to the nature of the wireless communication medium, there are certain advantages to implementing a WLL rather than a mobile system. These advantages emanate from the following reasons:

- A WLL has a fixed-to-fixed propagation path. The path loss of the fixed-to-fixed propagation in a WLL was based on 20 dB/decade (propagation path-slope, $g = 2$) in [34]. The path loss of the fixed-to-mobile propagation (i.e. cellular system) is often based on 40 dB/decade ($g = 4$). The path loss exponent is not dependent on the speed of either end of the link. Typically, the path loss exponent g lies between 3 and 5.
- In a WLL, the antennas are usually located at high spots. This implies that in a WLL system the received signal experiences less fading than in the fixed-to-mobile condition. The design E_b/N_0 ratio under the fixed-to-fixed condition for a 30 kHz channel is

assumed to be 14 dB, whereas the design E_b/N_0 ratio under a mobile radio condition for a 30 kHz channel is 18 dB.

- In a WLL, the frequency re-use distance can be reduced because in a WLL the fixed-to-fixed link may use directional antennas on both ends, so the interference area becomes small. A reduction in frequency re-use distance may provide an increase in the capacity.
- In a WLL, the use of adaptive antenna arrays on both the base and user stations does also increase the capacity making efficient use of spectrum [35].
- In a WLL, no handoffs occur because it is a fixed-to-fixed link. Also, the air link from each building to the cell site can be customarily installed to reduce the interference. Since the link remains unchanged (provided there is not too much growth and/or cell splitting) after installation, the design of the WLL system is much simpler than that of a mobile system.
- Since the WLL signal channel is a Gaussian noise channel or strong Rician channel (not a Rayleigh fading channel); no interleaving of the data stream is needed. Thus, data compression schemes and quadrature amplitude modulation (QAM) can be applied efficiently to generate a high rate throughout, although the radio transmission rate is as low as 14.4 kbps.

Appendix B   CDMA versus TDMA in WLL

Reviewing W. Webb's book [29], which constitutes an introduction to WLL, it is easily concluded that the first WLL systems were narrowband cellular systems properly adapted to the fixed subscriber conditions. CDMA and TDMA were the only multiple access techniques supporting the various air-interfaces. For this reason, the author has tried a detailed comparison of these two techniques in several aspects in order to distinguish the most efficient. It is interesting to see these aspects one by one.

1. Range   It is claimed that CDMA systems have a greater range than equivalent TDMA systems. Range is related to the path loss and the minimum signal level that the receiver can decode reliably. Path loss is independent of multiple-access methods, so the claim is basically one that CDMA can work with lower received signal strength than TDMA. This is true, since the receiver applies a gain, G, to the received signal with a CDMA network but not with a TDMA network. However, to counteract that, TDMA systems need only operate at a higher power level. Such an option is not possible with cellular systems because high power levels rapidly drain mobile batteries; power, however, is rarely a concern for a WLL operator. There seems little reason to support the claim that CDMA systems must have a greater range than TDMA systems.

2. Sectorization   Sectorization is the division of a circular cell into a number of wedge-shaped sectors. It is claimed that if sectorization is performed in a CDMA network, the number of sectors deployed can use the same frequency in each sector, increasing the capacity of the system. That claim is correct. It also is claimed that using sectors in a TDMA arrangement does not increase capacity. That also is broadly correct. Fundamentally, when a cell is sectorized, the cell radius remains the same. Hence, the transmitted power remains the same, and the distance, that is required to the next cell using the same frequency also remains the same. However, because there are now more cells within the sterilization radius, more frequencies, need to be found to avoid interference. So although the sector is smaller than the cell and thus has to support less traffic, it also has fewer frequencies on which to do that (because the total frequency assignment has been divided by a larger cluster size). That is not the case with CDMA, where using the same frequency in adjacent sectors increases only slightly the interference to neighbouring cells and slightly reduces their capacity.

TDMA could achieve a real gain if, instead of dividing a cell into a number of smaller cells by sectorization, the cell was divided into smaller circular cells, that is, the base stations were distributed around the cell and transmitted on a lower power level. That approach results in similar equipment costs but much higher site rental and backhaul costs; thus, it tends to be avoided except where absolutely necessary. In summary, the CDMA capacity can he increased by a factor of 2 to 3 by sectorization with only a small increase in cost. The option is not available in TDMA and hence represents an advantage of CDMA.

3. Frequency Planning   When different frequencies need to be assigned to neighbouring cells, the network planners have to decide which frequencies to use in which cells. In a CDMA system, where each frequency is used in each cell, no such decision needs to be made. For that reason, it is true that, in general, CDMA does not require frequency planning, although it may require PN code assignment planning. However, that is not a major advantage. Frequency planning can be readily accomplished with today's planning tools and easily adjusted if problems occur. DCA systems do not require frequency planning in any case. Finally, some CDMA WLL systems suggest that frequency planning be performed on a cluster size of 2, for various design reasons, so some frequency planning is required. In summary, frequency planning is not a key issue in the selection process for CDMA.

4. Operation in Unlicensed Bands   A number of frequency bands are unlicensed, that is, anyone can use them without having to obtain a license from the regulator. Such bands typically are used by *industrial, scientific, and medical* (ISM) applications, for example, ovens that use radio for heating purposes. Operating WLL systems in those bands has the single attraction that the spectrum is free and that the need to apply for a license is removed. However, the WLL system will suffer unknown and variable interference from uncontrolled sources.

These bands can be used only if the system can tolerate the interference. CDMA implicitly tolerates interference anyway, so in such an environment a CDMA system will work, but with reduced capacity. 'I'DMA systems cannot accommodate such inter-ference, implicitly, but there are techniques that allow them to do so. DCA selects channels according to the interference present on them. Another technique, frequency hopping, moves rapidly from channel to channel, so interference on one channel causes errors only for a short period of time.

CDMA systems cope slightly well with interference because they still use interfered channels but at a lower capacity, whereas TDMA systems use techniques to avoid transmitting on the channels. The capacity of a CDMA system probably is higher, then, in such an environment.

5. Macrocells versus Microcells   By using small cells in high-density areas, there are situations where smaller WLL cells are deployed within the coverage area of larger WLL cells. That is a problem for CDMA systems. Subscriber units configured for the larger cell will operate with much higher powers than those configured for the smaller cells. If both cells operated on the same frequency, the capacity of the smaller cell would be near zero, so different frequencies must be used. Because of the wide bandwidth and hence high-capacity of a CDMA system, that may be inefficient, in the worst case reducing the equivalent CDMA capacity by a factor of 2. Such reduction does not occur in TDMA, because the cells would be assigned different frequencies in any case.

The actual effect of microcells will vary from network to network, but with good planning, capacity reductions of far less than 2 should be realisable.

6. Risk TDMA systems have been widely deployed around the globe, while CDMA systems are far behind in number of built-up systems. There is a much higher risk with CDMA that equipment will be delayed, will not provide the promised capacity, will prove difficult to frequency plan, and so on. Such risks are continually reducing as experience with CDMA systems grows rapidly. Till now, however, the risk seems to have disappeared almost completely and the technology has matured very much among the manufacturers. Operators around the globe have started to take on such risks by establishing networks based on the CDMA air-interface as this has progressed all these years.

7. Cost Everything eventually comes down to cost. At the moment, CDMA system components cost more than TDMA system components. However, because of the higher capacity of CDMA systems, fewer base stations are required, resulting in lower equipment bills and lower site and line-rental costs. How the two facts balance depends on the actual difference in equipment costs and the extent to which the network is capacity limited. Certainly, in a highly capacity-limited situation, CDMA systems should prove less expensive.

8. Bandwidth Flexibility CDMA systems can increase the user bandwidth simply by reducing the gain factor G. TDMA systems also can be bandwidth flexible by assigning more than one TDMA slot per frame to a user. For example, DECT systems can assign between 32 and 552 Kbps dynamically to one user, depending on the load. Thus, both access methods are inherently, equally flexible although manufacturers may not have designed the capability into individual systems.

References

[1] V. K. Garg and E. L. Sneed, 'Digital Wireless Local Loop System,' *IEEE Commun. Mag.*, vol. 34, no. 10, pp. 112–115, Oct. 1996.

[1] B. Khasnabish, 'Broadband to the Home (BTTH): Architectures, Access Methods, and the Appetite for it,' *IEEE Network*, vol. 11, no. 1, pp. 58–69, Jan./Feb. 1997.

[2] H. Huh, S. C. Han and D. G. Jeong, 'WLL Services using Cellular Mobile Network,' Shinsegi Telecomm R&D/TR P2-97-04-02, Apr. 1997.

[3] H. Salgado, 'Spectrum Allocation for Fixed Wireless Access Technologies in the Americas,' in *Proc. IEEE VTC '98*, Ottawa, Canada, pp. 282–187, May 1998.

[4] A. R. Noerpel and Y. -B. Lin, 'Wireless Local Loop: Architecture, Technologies and Services,' *IEEE Personal Commun.*, vol. 5, no. 3, pp. 74–80, June 1998.

[5] W. C. Y. Lee, 'Spectrum and Technology of a Wireless Local Loop System,' *IEEE Personal Commun.*, vol. 5, no. 1, pp. 49–54, Feb. 1998.

[6] C. R. Baugh, E. Laborde, V. Pandey and V. Varma, 'Personal Access Communications System: Fixed Wireless Local Loop and Mobile Configuration and Services,' *Proc. IEEE*, vol. 86, no. 7, pp. 1498–1506, July 1998.

[7] C. C. Yu, D. Morton, C. Stumpf, R. G. White, J. E. Wilkes and M. Ulema, 'Low-Tier Wireless Local Loop Radio Systems, Part 1: Introduction,' *IEEE Commun. Mag.*, Mar. 1997.

[8] C. C. Yu, D. Morton, C. Stumpf, R. G. White, J. E. Wilkes and M. Ulema, 'Low-Tier Wireless Local Loop Radio Systems, Part 2: Comparison of Systems,' _IEEE Commun. Mag._, Mar. 1997.

[9] Z. I. Kostic and S. Seetharaman, 'Digital Signal Processors in Cellular Radio Communications,' _IEEE Commun. Mag._, Dec. 1997.

[10] M. P. Lotter and P. van Rooyen, 'CDMA and DECT: Alternative Wireless Local Loop Technologies for Developing Countries,' in _Proc. IEEE PIMRC '97_, Helsinki, Finland, pp. 169–173, Sep. 1997.

[11] TIA/EIA/IS-95-A, Mobile Station-Base Station Compatibility Standard for Dual-Mode Wideband Spread Spectrum Cellular System, 1995.

[12] ANSI J-STD-008, Personal Station-Base Station Compatibility Requirements for 1.8 to 2.0 GHz Code Division Multiple Access (CDMA) Personal Communications Systems, 1996.

[13] TIA/EIA/SP-3693 (to be published as TIA/EIA-95), Mobile Station-Base Station Compatibility Standard for Dual-Mode Wideband Spread Spectrum Cellular System, Baseline Version, July 1997.

[14] D. N. Knisely, S. Kumar, S. Laha and S. Nanda, 'Evolution of Wireless Data Services: IS-95 to cdma2000,' _IEEE Commun. Mag._, vol. 36, no. 10, pp. 140–149, Oct. 1998.

[15] D. G. Jeong and W. S. Jeon 'CDMA/TDD System for Wireless Multimedia Services with Traffic Unbalance between Uplink and Downlink,' _IEEE J. Select. Areas Commun._, vol. 17, no. 5, pp. 939–946, May 1999.

[16] ETSI/SMG/SMG2, The ETSI UMTS Terrestrial Radio Access (UTRA) ITU-R RTT Candidate Submission, Jan. 1998.

[17] Ad-hoc T, IMT-2000 Study Committee of ARIB, Japan's Proposal for Candidate Radio Transmission Technology on IMT-2000: W-CDMA, June 1998.

[18] TIA TR-45.5, The cdma2000 ITU-R RTT candidate submission, June 1998.

[19] TR41.6/96-03-007, Personal Wireless Telecommunications—Enhanced (PWT-E) Interoperability Standard (PWT-E), July 1996.

[20] ETS 300 175, Digital Enhanced Cordless Telecommunications (DECT); Common Interface (CI), European Telecommunications Standards Institute (ETSI), 1992.

[21] D. Akerberg, F. Brouwer, P. H. G. van de Berg and J. Jager, 'DECT Technology for Radio in the Local Loop,' in _Proc. IEEE VTC '94_, Stockholm, Sweden, pp. 1069–1073, June 1994.

[22] M. Gagnaire, 'An Overview of Broad-band Access Technologies,' _Proc. IEEE_, vol. 85, no. 12, pp. 1958–1972, Dec. 1997.

[23] M. Zanichelli, 'Cordless in the Local Loop,' in _Cordless Telecommunications in Europe_, W. Tuttlebee, Ed., Springer-Verlag, London, 1997.

[24] ANSI J-STD-014, Personal Access Communications Systems Air Interface Standard, 1995.

[25] A. R. Noerpel, Y. B. Lin and H. Sherry, 'PACS: Personal Access Communications System—A Tutorial,' _IEEE Personal. Commun._, pp. 32–43, June 1996.

[26] A. R. Noerpel, 'WLL: Wireless Local Loop—Alternative Technologies,' in _Proc. IEEE PIMRC_, Helsinki, Finland, 1997.

[27] R. C. Schulz, 'Wireless Technology and PCS Applications,' _GLA Int'l_, 1996.

[28] W. Webb, _Introduction to Wireless Local Loop_, Artech House, London, 1998.

[29] G. M. Stamatelos and D. D. Falconer, 'Milimeter Radio Access to Multimedia Services via LMDS,' in _Proc. IEEE GLOBECOM '96_, London, UK, pp. 1603–1607, Nov. 1996.

[30] W. Honcharenko, J. P. Kruys, D. Y. Lee and N. J. Shah 'Broadband Wireless Access,' _IEEE Commun. Mag._, Jan. 1997.

[31] J. Zhuang and M. E. Rollins, 'Forward Link Capacity for Integrated Voice/Data Traffic in CDMA Wireless Local Loops,' in _Proc. IEEE VTC '98_, Ottawa, Canada, pp. 1578–1582, May 1998.

[32] 'TES Quantum,' Tech. Rep. 1022896-0001, rev. B, Hughes Network Systems, 1996.

[33] W. C. Y. Lee, 'Applying CDMA to the Wireless Local Loop,' _Cellular Business_, pp. 78–86, Oct. 1995.

[34] A. G. Burr, 'A Spatial Channel Model to Evaluate the Influence of Directional Antennas in Broadband Radio System,' _IEE Colloquium on Broadband Digital Radio_, 1998.

2

Propagation Models for Wireless Local Loops

Dongsoo Har and Howard H. Xia

2.1 Introduction

Due to faster deployment and lower cost of *wireless local loop* (WLL) infrastructure as compared to a wired one, worldwide roll-out of WLL service has been highly anticipated. Most of WLL systems deployed so far belong to narrowband systems mainly aimed at providing voice service. These systems can be used as a bypass of wire-line local loop in dense areas and as an extension of existing telephone network in remote areas. In recent years, media-rich content of Internet has put speed pressure on the local loop. Application of WLL systems has been extended to broadband services to meet the need, contending with ISDN, *Asymmetrical Digital Subscriber Line* (ADSL) and cable TV. It is critical to understand the propagation characteristics of radio signal in the WLL environment to improve system economies of WLL services.

In order to predict path loss in wireless systems, signal variation over distance is typically expressed in terms of an inverse power law with a statistical shadowing component, that is obtained after averaging out the fast-fading effects. Specifically, the radio signal received at a receiver from a base station at a distance R can be written down as $10^{(\beta/10)}/R^{\gamma}$, where η represents the shadow effect and γ is the path loss exponent. In typical land-mobile radio environments, η is found to be a zero-mean Gaussian random variable with a standard deviation of 8 dB. Range dependence of path loss can also be expressed as an intercept–slope relationship in dB scale as

$$L \text{ (dB)} = I_1 + 10\gamma \log R \tag{2.1}$$

where I_1 is an intercept taken at a unit distance.

The size of a cell, in general, varies according to propagation environment and traffic density. Macrocell path loss models [1–4] are typically used for large cells with low traffic density. The prediction models [2,5–10] for small or medium cells are more appropriate for areas having moderate or high traffic density. Macrocell propagation models predict signal variations based on environment type, terrain variation, and morphology type (land use) rather than detailed environmental features such as building height and street width that are used for microcell models. Due to these higher resolution information used for prediction, microcell models generally provide more accurate prediction than macrocell models.

While radio signal can travel from a base station antenna to a receiving antenna via various paths, we assume here that a primary propagation path takes place over the rooftops in a building environment. Theoretical models dealing with such a propagation mechanism include *Walfisch–Bertoni* (WB) *model* [2], *COST 231-Walfisch–Ikegami* (COST 231-WI) *model* [5], *Xia–Bertoni* (XB) *model* [6], Vogler model [7], flat edge model [8], and slope diffraction model [9]. Empirical models representing such approaches are *Har–Xia–Bertoni* (HXB) *model* [10] and COST 231-WI model. These models are most appropriate in predicting path loss variation along non-LOS paths in low building environments.

Most path loss models only engage limited calculation of building reflection to reduce computation time. For example, reflections at buildings near the base station are normally neglected even though it is important for low antennas below the rooftop level. Numerical or recursive models such as the Vogler model, the flat edge model, and the slope diffraction model give complete representation of path loss for irregular heights and spacings of buildings. However, these numerical or recursive models are not convenient for analysis due to intensive computations required as number of building rows increases. In the rest of the small cell models, building environment along the primary signal path from base station to receiver is represented just by an average height when calculating path loss which are more appropriate for prediction of radio propagation in typical environments of buildings having quasi-uniform heights.

Subscriber antenna in WLL systems is typically fixed and placed on or around rooftop. On the other hand, most theoretical and empirical path loss models for macro- and micro-cells are applicable only for receivers well below surrounding rooftop level. To be used for prediction in WLL systems these path loss models need to be modified so that we can completely predict variation of received signal according to receiver location relative to rooftop. In this chapter, path loss models for WLL systems will be presented for receiving antenna heights ranging from 'on rooftop' to 'below rooftop'.

2.2 WLL System Configuration

WLL services can be classified into the following two categories:

- *Narrowband system.* The narrowband systems are typically used as an alternative to basic telephone services. Most of the WLL systems deployed so far belong to this category. This type of system provides voice service with limited support for data communication. Data rate available for this service is usually limited to several tens of kilo bits per second (Kbps). The system is mostly based on the existing cellular/PCS technologies with circuit switched connection.
- *Broadband system.* The broadband systems are intended to bypass wire-line local loop by providing high-speed, interactive services. Emerging broadband systems will be capable of supporting various services such as voice, high-speed Internet access and video-on-demand. Data rate required for these services can be up to several tens of giga bits per second (Gbps). Allocation of radio resource can be dynamic. The broadband network is anticipated to be packet-switched with guaranteed QoS.

In order to provide such services, a typical WLL system configuration which consists of wireless base station, subscriber unit and backbone switching network is shown in Figure 2.1. Base stations are interconnected through switching network by wire lines or microwave

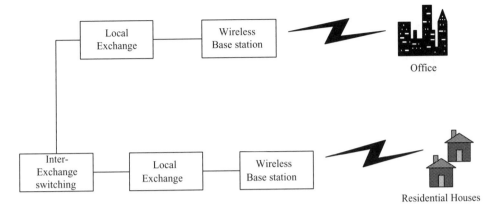

Figure 2.1 Configuration of wireless local loop

links. A subscriber unit generally consists of an antenna, a *network interface card* (NIC) and a subscriber device (usually a telephone). Because of the absence of definitive WLL radio standards, WLL systems can be implemented with the various radio technologies ranging from analogue to digital cellular, like AMPS, IS-95 CDMA and IS-136 TDMA and low-tier PCS such as *Cordless Telephone-2* (CT-2), *Digital Enhanced Cordless Telecommunications* (DECT) to proprietary systems.

2.3 Delay Spread in WLL Environments

Various propagation paths resulting from reflections at building walls and diffractions at building corners cause multipath fading. Because of the high transmitting power and large coverage area of macrocells, range of excess delays of significant multipath signal components is up to $10\,\mu s$ [11–12].

Power weighted average delay [13] is given by

$$\mu = \frac{\int tP(t)\mathrm{d}t}{\int P(t)\mathrm{d}t} \tag{2.2}$$

where t is delay parameter and $P(t)$ is referred to as power delay profile [11] representing the average power in the channel impulse response at t. From the average delay in Equation (2.2), second moment of delay parameter $t\sigma_d$ is defined as

$$\sigma_d = \sqrt{\frac{\int (t-\mu)^2 P(t)\mathrm{d}t}{\int P(t)\mathrm{d}t}} \tag{2.3}$$

σ_d is commonly referred to as RMS delay spread. A threshold can be used to eliminate insignificant multipath components at long delays. Typically, delayed signals that have powers greater than 25 or $30\,\mathrm{dB}$ below the peak response are only considered. Each envelope value of delayed signals are normalized by the signal mean over a small area or distance, removing the influence of received signal variations due to changes in distance from the transmitter [14].

Figure 2.2 shows examples of impulse response of radio channels. Figure 2.2(a) and 2.2(b) are profiles of impulse response for a macrocell channel with an elevated base station antenna and a microcell channel with low base station antenna of several meters above ground level. The macrocell channel in Figure 2.2(a) is found to have more multipath components of significant signal level relative to peak value as compared with the microcell channel in Figure 2.2(b). Cumulative distribution of received signal level is closer to Rayleigh distribution in case of macrocell whereas it is matched better to Ricean distribution with microcell channel.

It is found in [14] that, for 910 MHz, RMS delay spread of microcell channel computed with significant multipath components having power level greater than −25 dB with respect to the peak can be reduced by a factor of 4 as compared with the macrocell channel. Based on measurements at 1.9 GHz in a suburban area of St. Louis (US) with base station antenna at heights about the rooftop level of two story

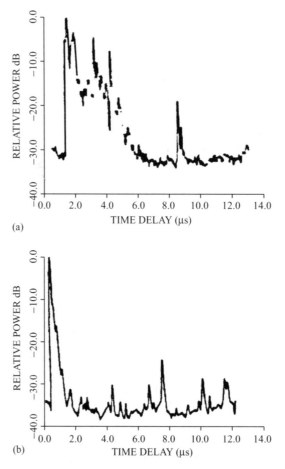

(a)

(b)

Figure 2.2 Examples of impulse response of radio channels: (a) macrocell (adopted from [11]), and (b) microcell (adopted from [14])

houses and subscriber unit at heights $2\sim3$ m, RMS delay spread doubled, statistically, for every 19 dB increment of path loss over a distance range less than 600 m [15]. Similar relation between delay spread and path loss was also observed in the microcell measurements [16] with low antennas ranging from 3 to 13 m. An upper bound of RMS delay spread was obtained as a function of path loss in [16]. It is expressed as

$$\sigma_d = \exp\left[0.065 * \text{PL}\right] \tag{2.4}$$

where σ_d is the RMS delay spread in nanoseconds and PL is the path loss in dB. For a receiving antenna located at top of a building and corresponding transmitters at top of nearby buildings, the measurements [17] at 1.9 GHz in urban environment of Madrid (Spain) resulted in delay spread 59.1, 54.9, 65.5 ns for antenna separation of 50, 150, 300 m with secured line-of-sight between transmitter and receiver.

2.4   Components of Overall Path Loss

As previously mentioned, current propagation models for small cells must be modified so that they can be applied for WLL planning. In this section, we will adjust the models, depending on locations of antennas, for WLL applications. For complete representation of path loss expression with various antenna locations, three cases are examined in detail. All the path loss expressions in this chapter are only for forward link (from base station to subscriber antenna). Path loss of reverse link can be obtained accordingly via the application of reciprocity principle.

Propagation models discussed in this chapter provide path loss value as a result of propagation over buildings and streets. Building and street parameters involved in the path loss calculation are average height of intervening buildings and average spacing of neighboured building rows. Among the models discussed in this chapter, HXB model does not explicitly include average spacing of building rows.

The average height of surrounding rooftops h_{BD} shown in Figure 2.3 can be used to determine relative antenna heights Δh_b and Δh_r. Specifically

$$\Delta h_b = h_b - h_{\text{BD}} \tag{2.5}$$

$$\Delta h_r = h_{\text{BD}} - h_r \tag{2.6}$$

where
   h_b = transmitting antenna height in meters
   h_{BD} = average building height in meters
   h_r = receiving antenna height in meters

Overall path loss PL in dB can be approximated [2,18] by the summation of (1) free space loss L_0, (2) loss due to intervening buildings L_{msd} and (3) loss due to diffraction at the last rooftop L_{rts}

$$\text{PL} = L_0 + L_{\text{msd}} + L_{\text{rts}} \tag{2.7}$$

While the mechanisms of two propagation processes associated with L_0 and L_{rts} are well understood and can be represented by the simple formulas, the multiple forward

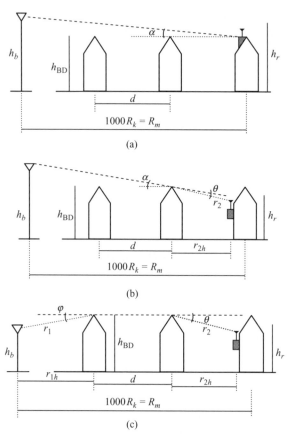

Figure 2.3 Propagation path in urban residential environment from base station to receiving antenna unit: (a) both antenna heights above the rooftop height h_{BD}, (b) only receiving antenna height is below rooftop height, and (c) both antenna heights below rooftop height

diffraction process pertinent to L_{msd} is not as simple since diffraction at edges of screens occurs in the transition region of previous screen. In order to account for various receiving antenna locations, modification of the propagation models is mainly involved with L_{rts}, particularly the parameters, Δh_b and Δh_r.

2.4.1   Free Space Loss L_0

Free space loss accounts for the signal attenuation due to spherical spreading of the wavefront excited by a point source. The free space loss incurred between isotropic antennas of transmitter and receiver is given by

$$L_0 = -10\log\left(\frac{\lambda}{4\pi R_m}\right)^2 \qquad (2.8)$$

where

λ is the wavelength in meters

R_m is the separation between transmitting and receiving antennas in meters.

Alternatively, Equation (2.8) can be expressed in dB as a function of distance and frequency

$$L_0 = 32.4 + 20 \log R_k + 20 \log f_M \tag{2.9}$$

where

R_k = antenna separation in km

f_M = frequecny in MHz

2.4.2 Loss due to Multiple Forward Diffractions Passing Intervening Rooftops L_{msd}

In order to find the effect of intervening buildings between base station and receiver Walfisch and Bertoni [2] evaluated numerically the reduction of the field for incident plane wave passing through multiple screens for base station antenna above surrounding rooftops. Following this study, Xia and Bertoni [6] provided theoretical field reduction in cases of incident cylindrical and plane waves. The use of XB model is also valid for base station antenna below surrounding rooftops. Results of Xia and Bertoni [6] confirmed those calculated using the plane wave approach in [2] for base station antennas above the rooftops.

The centre-to-centre spacing of building rows is typically of the order of 50 m. Path loss is often predicted up to several kilometers. As a result, intervening buildings between base station and receiver can be simplified by an array of absorbing screens as seen in Figure 2.4 in evaluating the signal level at receiver location of interest. With base station antenna a few meters above rooftop level, glancing angle α in Figure 2.3(a) will be small. For small glancing angle of incident wave, certain degree of irregularities of building height, spacing and lack of parallelism of building rows have little effect on overall path loss, so average spacing of buildings of average height can be applied for path loss prediction [19]. Received field at the M-th rooftop shown in Figure 2.4, which is the nearest rooftop to the receiver location of interest, has a loss L_{msd} as a result of diffractions by $M - 1$ screens. The reduction of the field can be expressed by a factor Q which is a function of dimensionless propagation parameter [2]. Using this factor, L_{msd} is given by

$$L_{msd} = -10 \log Q^2 \tag{2.10}$$

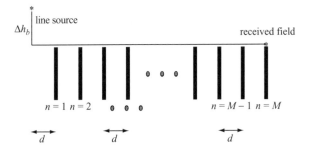

Figure 2.4 A series of thin absorbing half screens replacing buildings for path loss prediction (adopted from [25])

2.4.3   Loss due to Diffraction at the Rooftop Nearest to Receiver L_{rts}

Path loss associated with diffraction down to street level depends on the shape and configuration of buildings in the vicinity of the receiver. Using the *Geometrical Theory of Diffraction* (GTD) [20], loss due to this diffraction at the rooftop nearest to receiver shown in Figure 2.5, L_{rts} for small glancing angle α close to 0, is obtained as

$$L_{rts} = -10\log\left(\frac{1}{2\pi k r_1 \theta^2}\right) \tag{2.11}$$

where

$$r_1 = \sqrt{r_h^2 + \Delta h_r^2}$$

$$\theta \approx \tan^{-1}\left(\frac{\Delta h_r}{r_h}\right)$$

$$k = \text{wave vector} = \frac{2\pi}{\lambda}$$

For small θ, $r_1 \approx r_h$ and $\theta \approx (\Delta h_r/r_h)$. Reflections from the building next to the mobile and other multipath signals result in doubling the amplitude of the field reaching receiver directly from the last rooftop. To take the reflected signals into account, a factor 2 can be inserted inside the bracket of Equation (2.11) [2,18]. Similar factors have been applied to predict FM radio and TV signal strength at ground level. Based on an empirical model developed by the US *Environment Protection Agency* (EPA), ground reflection leads to a maximum increase of signal strength of 2.56 [21]. With a factor 2 accounting for reflection from a building next to receiver located at a distance $(1/2)d$ from the last rooftop, (6-A) can be rewritten as

$$L_{rts} = 21.8 - 10\log d + 10\log f_G + 20\log \Delta h_r \tag{2.12}$$

At a transition region where $\theta \approx 0$ radian, L_{rts} given in Equation (2.11) has unbounded value. A transition function F is needed to remove the singularity. With the inclusion of transition function, rooftop-to-receiver loss L_{rts} is given by [22–23]

$$L_{rts} \approx -10\log\left(|2F(s)|^2 \cdot \frac{1}{2\pi k r_1 \theta^2}\right) \tag{2.13}$$

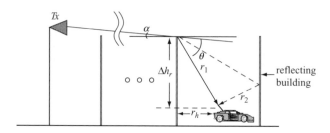

Figure 2.5   Geometry for L_{rts}

where

$$s = \frac{k\Delta h_r^2}{2r_h} = \frac{kr_h}{2}\left(\frac{\Delta h_r}{r_h}\right)^2$$

$$F(s) = \text{transition function} = \sqrt{2\pi s}\left[f\left(\sqrt{\frac{2s}{\pi}}\right) + jg\left(\sqrt{\frac{2s}{\pi}}\right)\right]$$

The functions $f(x)$ and $g(x)$ can be obtained from the following rational approximations [24]:

$$f(x) = \frac{1 + 0.926x}{2 + 1.792x + 3.104x^2}, \quad g(x) = \frac{1}{2 + 4.142x + 3.492x^2 + 6.670x^3} \tag{2.14}$$

Near the shadow boundary where $s \ll 1$, $f(x)$ and $g(x)$ in (6-D) are close to 1/2 so that $|F(s)| = \sqrt{\pi s}$. Substituting $\sqrt{\pi s}$, $\Delta h_r/r_h$ for $|F(s)|$, θ in (6-C), $L_{rts} = 0$ for $\Delta h_r = 0$ so that L_{rts} is continuous for the range of the receiving antenna height $\Delta h_r \le 0$. Note that the factor 2 of the term $2F(s)$ ensures L_{rts} to be 0 when $\Delta h_r = 0$ and is not representing the impact of reflection via the ray associated with r_2 in Figure 2.5.

Figure 2.6 shows a comparison of L_{rts} based on Equations (2.11) and (2.13) for various diffraction angles θ. The factor 2 accounting for the reflections from a building next to the mobile is inserted into Equation (2.11). From Figure 2.6, it is seen that L_{rts} based on Equation (2.13) is, as expected, 0 when $\theta = 0$ radian and there is 3 dB difference for $\theta > 0.1$ radian between the values based on two different L_{rts} evaluations.

2.5   Path Loss Models

Generally, path loss model is valid for a specific range of base station antenna heights, building heights, frequency, and antenna separation. In this section, path loss models are

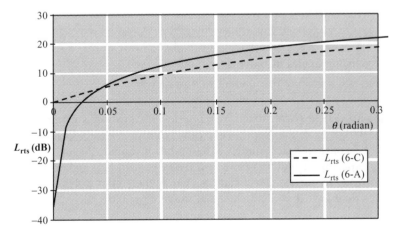

Figure 2.6   Comparison L_{rts} values from (6-A) and (6-C)

classified in terms of height of base station relative to surrounding buildings as well as the height of receiving antenna relative to the building in the vicinity of the receiver. Modification of the path loss models is carried out according to antenna heights relative to surrounding buildings.

2.5.1 Both the Transmitting and Receiving Antennas are above Rooftop Level ($\Delta h_b > 0$ and $\Delta h_r > 0$)

For multiple diffractions passing the absorbing half screens the field at the edge of each screen can be obtained from the numerical evaluation via repeated Kirchhoff–Huygens integral. Since diffraction process in the vertical plane that contains base station antenna, receiver and edges of screens is the same as it is with the field excited by a point source local plane wave approximation was used to calculate the effect of the buildings. A frequency–angle dependence was found, meaning that curves having the same value of $\alpha(d/\lambda)^{1/2}$ ($\equiv g_p$), are approximately the same by an accuracy percentage less than two percent [2]. For example, the variation of magnetic field level at n-th edge H_n with $d = 200\,\lambda$ and $\alpha = 0.4°$ is comparable with that corresponding to $d = 50\,\lambda$ and $\alpha = 0.8°$. It is shown in Figure 2.7 that the field settles to a nearly constant value after an initial drop to a minimum for n large enough. In Figure 2.8, this behaviour is illustrated for $d = 50\,\lambda$ and $\alpha = 1.2°$. The field drops to a minimum and gradually increases as n increases. The screen number N_0 in Figure 2.8 corresponding to the edge for which amplitude of field has the settled value is shown for each value of α by the vertical stroke in Figure 2.7. From the relation between the field and g_p dependence of Q on the parameter g_p was obtained. Loss L_{msd} due to diffractions at multiple screens was obtained with a polynomial fit [2] given by

$$L_{msd} = -10 \log Q^2(g_p) \tag{2.15}$$

where

$$Q(g_p) = 2.35 \, g_p^{0.9}$$

$$g_p = \alpha \sqrt{\frac{d}{\lambda}} = \tan^{-1}\left(\frac{\Delta h_b}{R_m}\right)\sqrt{\frac{d}{\lambda}} \approx \frac{\Delta h_b}{R_m}\sqrt{\frac{d}{\lambda}}$$

$\alpha =$ glancing angles in radians

$d =$ average spacing of building row in meters

The fit to the settled field Q calculated by numerical integration is within 0.8 dB accuracy over the range $0.01 < g_p < 0.4$. Appropriate range of distance for which the valid range of g_p holds can be computed according to base station antena height, average building height, average spacing of building rows and frequency. In [25] higher-order polynomial fit was obtained to use for smaller distances. The higher-order polynomial fit having an accuracy better than 0.5 dB over an extended range of g_p $0.01 < g_p < 1.0$ is given by

$$Q(g_p) = 3.502 \, g_p - 3.327 \, g_p^2 + 0.962 \, g_p^3 \tag{2.16}$$

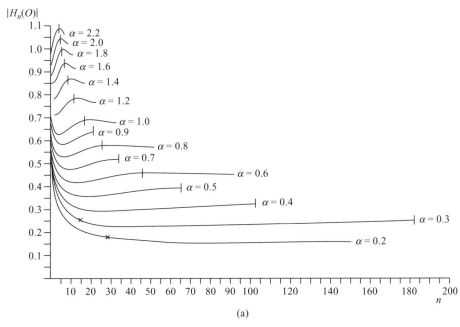

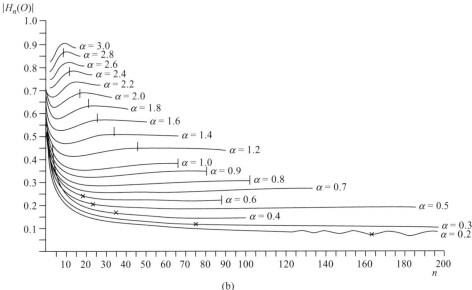

Figure 2.7 Variation of field incident on edges of the half screens as function of screen number n for various values of the glancing angle a for average screen spacing: (a) $d = 200\,\lambda$, and (b) $d = 50\,\lambda$ (adopted from [2])

Using Equations (2.8), (2.13), and (2.15), for a receiver equal to or below the rooftop level, the overall path loss is expressed as

$$\text{PL} = -10\log\left(\frac{\lambda}{4\pi R_m}\right)^2 - 10\log Q^2(g_p) - 10\log\left(|2F(s)|^2 \cdot \frac{1}{2\pi k r_2 \theta^2}\right) \qquad (2.17)$$

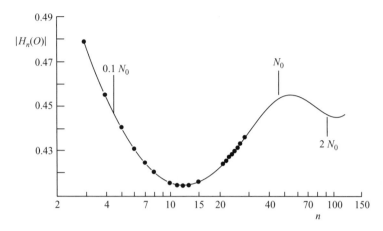

Figure 2.8   Settling behaviour of the field for $d = 50\lambda$ and $\alpha = 1.2°$ (adopted from [2])

For receiver location as shown in Figure 2.3(a), there is direct path connecting base station antenna and receiving antenna. Signal reaching the receiver is different from that travelling in free space since the buildings affect the wave spreading. As a function of distance above the rooftops, the field is seen to be in the form of standing wave with a peak-to-peak spacing $\lambda/(2\sin\alpha)$. Interference of standing wave can be accounted for by the incident plane wave, and a reflected plane wave propagating away from the edges at an angle α with regard to horizontal direction [26]. Since the field at the rooftop for a unit amplitude plane wave is $Q(g_p)$, the standing wave Q_s is of the form

$$Q_s = \exp(jk\Delta h_r \sin\alpha) + (Q(g_p) - 1)\exp(-jk\Delta h_r \sin\alpha)$$
$$= 2j\sin(k\Delta h_r \sin\alpha) + Q(g_p)\exp(-jk\Delta h_r \sin\alpha) \tag{2.18}$$

Hence squared-magnitude of Q_S can be approximated as

$$|Q_s|^2 = |+ 2j\sin(k\Delta h_r \sin\alpha) + Q(g_p)\exp(-jk\Delta h_r \sin\alpha)|^2$$
$$\approx Q^2(g_p) + 4(k\Delta h_r)^2 \sin^2\alpha \tag{2.19}$$

where

$$Q^2(g_p) \ll 1$$

$$k\Delta h_r \sin\alpha \ll 1$$

are assumed. The path loss accounting for free space loss L_0 combined with excess loss L_{msd} in case of receiving antenna mounted on rooftop can be obtained [27] as

$$\mathrm{PL} = -10\log\left(\frac{\lambda}{4\pi R_m}\right)^2 - 10\log\left\{Q^2(g_p) + 4[(2\pi/\lambda)\Delta h_r \sin\alpha]^2\right\} \tag{2.20}$$

Path loss model given by Equation (2.20) is valid for conditions, α (in degree) $\leq 2°$ and $(\Delta h_b \Delta h_r / \lambda R_m) < (1/8)$.

When height of receiving antenna is close to rooftop level, i.e. $\Delta h_r \approx 0$, bracketed term in Equation (2.20) is vanished, so Equation (2.20) becomes, as expected, WB model without rooftop-to-receiver diffraction term L_{rts}.

2.5.2 Receiving Antenna is below the Rooftop ($\Delta h_b > 0$ and $\Delta h_r < 0$)

When base station antenna is above the rooftop level and receiver is below or equal to rooftop level as shown in Figure 2.3(b), path loss models such as WB model, COST 231-WI model, and HXB model need to be modified to calculate the loss L_{rts} due to diffraction at the last rooftop to receiver.

2.5.2.1 Modification of COST 231-WI Model and WB Model

European *Co*operation in the field of *S*cientific and *T*echnical Research (COST) 231 group has developed outdoor path loss models for applications in urban areas at frequencies of the cellular and PCS bands. Based on extensive measurements performed in European cities, COST 231 group has modified various path loss models for microcellular environments. Two theoretical models WB model and Ikegami model [28] are combined with the results obtained from measurements made in various European cities to formulate the COST 231-WI model. The COST 231-WI model utilizes the theoretical WB model to obtain multiple screen forward diffraction loss L_{msd} for high antennas (above surrounding buildings) whereas it uses measurement-based L_{msd} for low antennas (below the buildings).

In case of path loss prediction for non-LOS routes overall path loss PL is composed of three terms, free space loss L_0, multiple screen diffraction loss L_{msd}, and rooftop-to-street diffraction loss L_{rts} in the form of

$$PL = \begin{cases} L_0 + L_{rts} + L_{msd} & \text{for } L_{rts} + L_{msd} > 0 \\ L_0 & \text{for } L_{rts} + L_{msd} < 0 \end{cases} \tag{2.21}$$

L_{rts} takes into account the width of the street and its orientation. With street orientation factor L_{ori}, it is given by

$$L_{rts} = -16.9 - 10 \log d + 10 \log f_M + 20 \log \Delta h_r + L_{ori} \tag{2.22}$$

where

$$L_{ori} = \begin{cases} -10 + 0.354\,\varphi & 0° \leq \varphi < 35° \\ 2.5 + 0.075\,(\varphi - 35) & 35° \leq \varphi < 55° \\ 4.0 - 0.114\,(\varphi - 55) & 55° \leq \varphi \leq 90° \end{cases}$$

Street orientation angle φ is illustrated in Figure 2.9. Excess loss L_{msd} of COST 231-WI model was obtained as

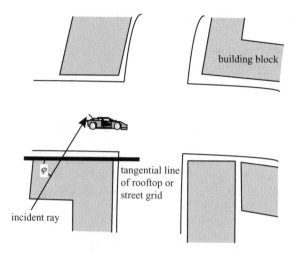

Figure 2.9 Pictorial definition of street orientation angle φ

$$L_{\text{msd}} = L_{\text{bsh}} + k_a + k_d \log R_k + k_f \log f_M - 9 \log d \qquad (2.23)$$

Each term in Equation (2.23) is given by

$$L_{\text{bsh}} = \begin{cases} -18 \log(1 + \Delta h_b) & \Delta h_b > 0 \\ 0 & \Delta h_b \leq 0 \end{cases}$$

$$k_a = \begin{cases} 54 & \Delta h_b > 0 \\ 54 - 0.8 \, \Delta h_b & R_k \geq 0.5 \text{ and } \Delta h_b \leq 0 \\ 54 - 0.8 \, \Delta h_b (R_k/0.5) & R_k < 0.5 \text{ and } \Delta h_b \leq 0 \end{cases}$$

$$k_d = \begin{cases} 18 & \Delta h_b < 0 \\ 18 - 15(\Delta h_b/h_{BD}) & \Delta h_b \leq 0 \end{cases}$$

$$k_f = -4 + \begin{cases} 0.7([f_M/925] - 1) & \text{medium sized cities and suburban centres} \\ & \text{with moderate tree density} \\ 1.5([f_M/925] - 1) & \text{metropolitan centres} \end{cases}$$

The term k_a indicates the increase of the path loss for base station antennas below the rooftops of the neighboured buildings. k_d and k_f represent the dependence of the multi-screen diffraction loss versus distance and the frequency, respectively. It is claimed that the estimation of path loss agrees rather well with measurements for base station antenna heights above rooftop level. Also the mean error is in the range of ± 3 dB and the standard deviation is in the range of 4–8 dB [5]. Overall path loss PL can be obtained by combining Equations (2.13), (2.22) and (2.23).

2.5.2.2 *Modified HXB Model*

This model is based on formulas representing the regression fits to measurements given in the 'Outdoor signal strength test' portion of the *Telesis Technologies Lab.* (TTL) report [29] to the FCC and published in [30–31]. The measurements were made in the cellular and PCS frequency bands, 0.9 and 1.9 GHz. In the measurements, base station height h_b is varied from 3.2 m to 8.7 m to 13.4 m and mobile antenna height h_r is fixed at 1.6 m. The Sunset District, and the Mission District of San Francisco city, which have attached buildings of quasi-uniform height built on a rectangular street grid on flat terrain, were selected as typical low-rise environments. Figure 2.10 shows the test routes for a transmitter located on the street in the middle of a block in a region characterized by a rectangular street grid. Measurements were performed for radial distances up to 3 km.

Signal strength on the zig-zag route showed 10–20 dB decreases as the mobile turned a corner from the perpendicular street into streets parallel to that of the base station [30]. As a result, the measurements for the two different segments of this path were treated as separate groups. On the parallel streets the propagation path is *transverse* to the rows of buildings. On the perpendicular streets the propagation path has a long *lateral* segment down the street. Signal strength on the *staircase* route showed continuous variation with distance traveled by the mobile, so that measurements were treated as one group.

Using the intercepts and slope indices obtained from the measurements, the path loss formulas were obtained [10] as

Staircase Route

$$\begin{aligned}
\text{PL}(R_k) = &[137.61 + 35.16 \log f_G] - [12.48 + 4.16 \log f_G] \, \text{sgn}(\Delta h_b) \log(1 + |\Delta h_b|) \\
&+ [39.46 - 4.13 \, \text{sgn}(\Delta h_b) \log(1 + |\Delta h_b|)] \log R_k
\end{aligned} \tag{2.24}$$

Transverse Route

$$\begin{aligned}
\text{PL}(R_k) = &[139.01 + 42.59 \log f_G] - [14.97 + 4.99 \log f_G] \, \text{sgn}(\Delta h_b) \log(1 + |\Delta h_b|) \\
&+ [40.67 - 4.57 \, \text{sgn}(\Delta h_b) \log(1 + |\Delta h_b|)] \log R_k
\end{aligned} \tag{2.25}$$

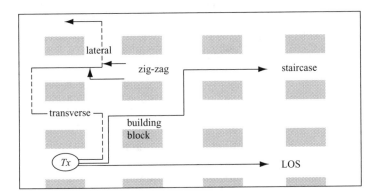

Figure 2.10 Staircase, zig-zag (transverse + lateral) and LOS test routes relative to street grid

Lateral Route

$$PL(R_k) = [127.39 + 31.63 \log f_G] - [13.05 + 4.35 \log f_G] \, \mathrm{sgn}(\Delta h_b) \log(1 + |\Delta h_b|)$$
$$+ [29.18 - 6.70 \, \mathrm{sgn}(\Delta h_b) \log(1 + |\Delta h_b|)] \log R_k \qquad (2.26)$$

where the sign function, $\mathrm{sgn}(x)$ is defined as

$$\mathrm{sgn}(x) = \begin{cases} +1 & x \geq 0 \\ -1 & \text{otherwise} \end{cases} \qquad (2.27)$$

and f_G is frequency in GHz. Note that the first constant on the right side in Equations (2.24)–(2.25) indicates path loss value at 1 km for 1 GHz with $\Delta h_b = 0$ m. Considering the range of the parameters over which measurements were made, these formulas are valid for $0.9 < f_G < 2$ GHz, $-8 < \Delta h_b < 6$ m, $R_k < 3$ km.

Since influence of the variation in building height and street width on L_{rts} is represented in terms of Δh_r and r_h, theoretical correction factors relevant to a specific building height and an arbitrary street width are a function of these parameters. The geometrical average building height in Sunset and Mission is 7.8 m relative to the mobile height of 1.6 m. Thus, the correction factor for a given building height is given by

$$(\Delta PL)_{\Delta h_r} = 20 \log(\Delta h_r / 7.8) \qquad (2.28)$$

It is seen in Figure 2.11 that path loss level corresponding to lateral route is significantly lower than the other non-LOS routes. For the receivers on non-LOS routes shown in Figure 2.12, each path to a receiver crosses rooftops that are represented by the edges of absorbing half screens placed at the middle of the buildings. To simplify the evaluation of diffraction over the rooftops, each absorbing screen at the diffraction point is oriented

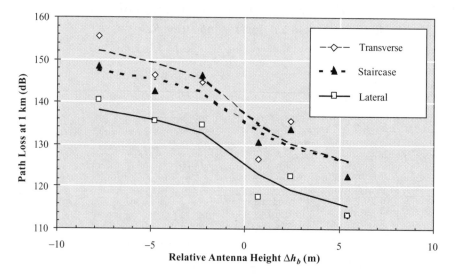

Figure 2.11   Dependence of 1 km path loss intercepts on Δh_b

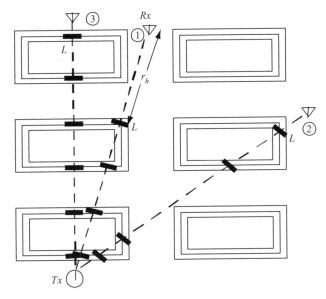

Figure 2.12 Simplified footprint of townhouses and ray paths associated with non-LOS routes

perpendicular to the direction of ray path [32], as indicated by the dark crossing lines in Figure 2.12. It is seen in Figure 2.12 that distance r_h from the last rooftop, which is marked 'L', to receiver ① on lateral route is large whereas r_h for receivers ②, ③ is relatively small. Also, the number of half screens, and the spacing between the screens, is seen to be different for different routes. However, path loss for base station antenna heights near to the rooftops is not sensitive to irregularities in the row spacing along the ray path. Moreover, loss L_{msd} varies as $20\log(M)$ [2,6], where M is the number of screens, and hence is not strongly dependent on M, especially near the cell boundary where M is large. Therefore, the large value of r_h can be regarded as the principal cause for the small path loss associated with the lateral route.

In order to reflect the effect of r_h on path loss level, a theoretical correction factor $(\Delta\mathrm{PL})_{r_h}$ based on the dependence on r_h can be used. Since the distance between the building fronts and the centre of the street is about 20 m in Sunset and Mission districts, using the transverse route formula as the standard formula, the correction factor $(\Delta\mathrm{PL})_{r_h}$ for other routes is obtained as

$$(\Delta\mathrm{PL})_{r_h} = 10\log(20/r_h) \tag{2.29}$$

From the foregoing discussion, an anisotropic formula which applies to all non-LOS routes by explicitly including the distance r_h is obtained as

All non-LOS Routes

$$\begin{aligned}
\mathrm{PL}(R_k) =& [139.01 + 42.59\log f_G] - [14.97 + 4.99\log f_G]\,\mathrm{sgn}(\Delta h)\log(1+|\Delta h|) \\
& + [40.67 - 4.57\,\mathrm{sgn}(\Delta h)\log(1+|\Delta h|)]\log R_k + 20\log(\Delta h_r/7.8) \\
& + 10\log(20/r_h)
\end{aligned} \tag{2.30}$$

Note that formula parameters Δh_r, r_h in Equation (2.30) cause unrealistic path loss values when they are close to $0\,m$. Since distance r_h for the staircase route is generally a little larger than that of transverse route, the non-LOS formula (2.30) will give lower path loss for the staircase route. The discrepancy between the two predictions based on lateral route formula in Equation (2.26) and non-LOS formula in Equation (2.30) can be shown to be about several dBs for a range of base station antenna height used for the measurements.

Since path loss expression in Equation (2.30) was modified from transverse route formula which is pertinent to receiver location ③, L_{rts} component in Equation (2.12) can be removed from Equation (2.30) by setting $d = 2r_h$. After the subtraction process, free space loss combined with multiple screen diffraction loss $L_0 + L_{msd}$ can be expressed as

$$L_0 + L_{msd} = [115.38 + 32.59 \log f_G] - [14.97 + 4.99 \log f_G] \operatorname{sgn}(\Delta h_b) \log(1 + |\Delta h_b|)$$
$$+ [40.67 - 4.57 \operatorname{sgn}(\Delta h_b) \log(1 + |\Delta h_b|)] \log R_k \qquad (2.31)$$

Hence, the total path loss based on HXB model is modified as

$$PL = (11 - G) + (6 - C) \qquad (2.32)$$

For receiving antennas mounted on rooftops, Equation (2.31) can be utilized further, to get L_{msd}. Furthermore, adjusting the unit of frequency in the expression (2.20) for free space loss, L_{msd} of HXB model can be obtained by subtracting expression (2.20) from (2.31)

$$L_{msd, HXB} \approx [23.00 + 12.59 \log f_G] - [14.97 + 4.99 \log f_G] \operatorname{sgn}(\Delta h_b) \log(1 + |\Delta h_b|)$$
$$+ [20.67 - 4.57 \operatorname{sgn}(\Delta h_b) \log(1 + |\Delta h_b|)] \log R_k \qquad (2.33)$$

With Equation (2.33) path loss of the receiving antennas above the rooftops can be expressed as

$$PL = -10 \log \left(\frac{\lambda}{4\pi R_m}\right)^2 - 10 \log \left\{ Q_{HXB}^2 + 4[(2\pi/\lambda)\Delta h_r \sin \alpha]^2 \right\} \qquad (2.34)$$

where

$$Q_{HXB}^2 = 10^{-0.1 L_{msd, HXB}}$$

2.5.3 Both the Transmitting and Receiving Antennas below the Rooftop ($\Delta h_b < 0$ and $\Delta h_r < 0$)

For a cylindrical wave excited by a line source with $M - 1$ absorbing half screens spaced d apart, the field reaching the edge of M-th screen was analytically evaluated by Xia and Bertoni [6]. It is shown in [6] that L_{msd} according to XB models for transmitting antenna height at rooftop level can be obtained as

$$L_{msd} = -10 \log Q_M^2 \qquad (2.35)$$

where

$$Q_M = \sqrt{M} \left| \sum_{q=0}^{\infty} \frac{1}{q!} \left[2g_c \sqrt{j\pi} \right]^q I_{M-1,q} \right|$$

$$g_c = \Delta h_b \frac{1}{\sqrt{\lambda d}}$$

The Boersma functions [33] are found from the following recursion relation:

$$I_{M-1,q} = \frac{(M-1)(q-1)}{2M} I_{M-1,q-2} + \frac{1}{2\sqrt{\pi M}} \sum_{n=1}^{M-2} \frac{I_{M-1,q-1}}{(M-1-n)^{1/2}} \qquad (2.36)$$

with initial terms

$$I_{M-1,0} = \frac{1}{M^{3/2}}$$

$$I_{M-1,1} = \frac{1}{4\sqrt{\pi}} \sum_{n=0}^{M-1} \frac{1}{n^{3/2}(M-n)^{3/2}}$$

Q_M in Equation (2.35) indicates that it depends on relative height of base station antenna to rooftops Δh_b and average spacing of building rows d through the parameter g_c. The results of Q_M calculations for $\Delta h_b > 0$ m, which is in good agreement with Q value based on WB model, were shown in [25]. When both the antenna heights are below the rooftop level as shown in Figure 2.3(c) simple approximation of path loss can be used for XB model.

When base station antenna is sufficiently below the rooftop, the second row of building lies outside the transition region of the first row of buildings and the process of multiple forward diffractions can be decomposed into two distinct wave diffraction processes. The cylindrical wave excited by a line source below the average rooftop is diffracted by the first row of buildings. The first row of buildings then acts approximately as a line source for the diffractions at the rest of buildings. The first process can be identically treated by GTD as with diffraction from the last rooftop to receiver while the latter multiple diffraction process has been evaluated in a closed form by Xia and Bertoni [6]. Under the conditions $\Delta h_b = 0$ and $R_m = Md$, Q_M in Equation (2.35) reduces to a simple closed-form solution

$$Q_M = \frac{1}{M} = \frac{d}{R_m} \qquad (2.37)$$

where M is the number of building rows between antennas. When a base station antenna is sufficiently below the rooftop the second row of buildings lies outside the transition region of the first row of buildings. For a cylindrical wave incident to edge of the first row of buildings at an angle $\varphi = \tan^{-1}(\Delta h_b/r_{1h})$, the field reduction due to the combined contributions of two cylindrical wave diffraction processes can be expressed [25] as

$$Q_M \approx \frac{1}{(M-1)} \frac{1}{\varphi\sqrt{2\pi k r_1}} \qquad \text{for } \Delta h_b < 0 \qquad (2.38)$$

where

$$r_1 = \sqrt{\Delta h_b^2 + r_{1h}^2} \text{ in meters}$$

The overall path loss in dB is then obtained based on XB model as the summation of Equations (2.8), (2.13) and either (2.37) or (2.38).

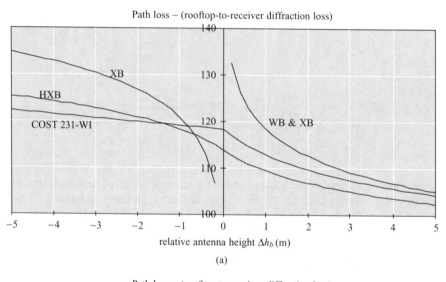

(a)

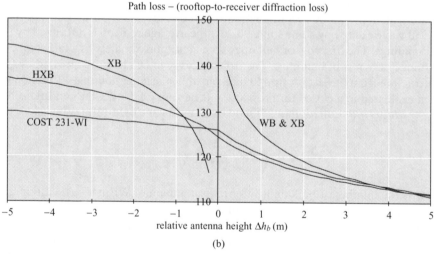

(b)

Figure 2.13 Comparison of path loss values excluding rooftop-to-receiver diffraction loss according to four path loss models for (a) 0.9 GHz, and (b) 1.9 GHz. Relevant parameters are antenna separation $R_k = 1$ km, average building height $h_{BD} = 8$ m, average spacing of building row $d = 50$ m, distance between base station antenna and first building row $r_{1h} = 50$ m (only for XB model)

2.6 Comparison of Propagation Models

Foregoing models for WLL system design are compared for relative antenna height in a range of $-5\,\mathrm{m} < \Delta h_b < 5\,\mathrm{m}$. In Figure 2.13, we have plotted combined loss $L_0 + L_{\mathrm{msd}}$ at a distance 1 km for frequencies of 0.9 GHz and 1.9 GHz. For loss L_{msd} for Δh_b range $-5\,\mathrm{m} < \Delta h_b < 0\,\mathrm{m}$, Equation (2.38) was used for XB model. It seems that theoretical models, WB model and XB model, are generally a little more pessimistic as compared to the empirical models, COST 231-WI model and HXB model. The singularity of WB model and XB model at $\Delta h_b = 0\,\mathrm{m}$ is due to the value 0 of Q factor in Equation (2.15) and the use of unbounded value of Q_M in Equation (2.38), respectively. Path loss difference between XB model and COST 231-WI model or HXB model increases as base station antenna height decreases, while the difference between WB model and the other two models increases as the base station antenna is getting closer to rooftop level.

References

[1] M. Hata, 'Empirical Formula for Propagation Loss in Land Mobile Radio Services,' *IEEE Trans. Veh. Technol.*, vol. 29, pp. 317–325, Aug. 1980.

[2] J. Walfisch and H. L. Bertoni, 'A Theoretical Model of UHF Propagation in Urban Environments,' *IEEE Trans. Antennas Propagat.*, vol. 36, pp. 1788–1796, 1988.

[3] K. Allsebrook and J. D. Parsons, 'Mobile Radio Propagation in British Cities at Frequencies in the UHF and VHF Bands,' *IEEE Proc.*, vol. 124, no. 2, pp. 95–102, 1977.

[4] W. C. Y. Lee, *Mobile Communications Engineering*, Chapter 3, McGraw Hill, New York, 1982.

[5] COST 231 Final Report; *Propagation Prediction Models*, Chapter 4, pp. 17–21.

[6] H. H. Xia and H. L. Bertoni, 'Diffraction of Cylindrical and Plane Waves by an Array of Absorbing Half Screens,' *IEEE Trans. Antennas Propagat.*, vol. 40, pp. 170–177, 1992.

[7] L. E. Vogler, 'An Attenuation Function for Multiple Knife-edge Diffraction,' *Radio Sci.*, vol. 17, pp. 1541–1546, 1982.

[8] S. R. Saunders and F. R. Bonar, 'Prediction of Mobile Radio Wave Propagation over Buildings of Irregular Heights and Spacings,' *IEEE Trans. Antennas Propagat.*, vol. 42, pp. 137–144, 1994.

[9] J. B. Anderson, 'UTD Multiple-Edge Transition Zone Diffraction,' *IEEE Trans. Antennas Propagat.*, vol. 45, pp. 1093–1097, 1997.

[10] D. Har, H. H. Xia and H. L. Bertoni, 'Path Loss Prediction Model for Microcells,' accepted for publication through *IEEE Trans. Veh. Technol.*

[11] D. C. Cox, 'Delay Doppler Characteritics of Multiple Propagation at 910 MHz in a Suburban Mobile Radio Environment,' *IEEE Trans. Antennas Propagat.*, vol. 20, Sep. 1972.

[12] ——, '910 MHz Urban Mobile Radio Propagation: Multipath Characteristics in New York City,' *IEEE Trans. Commun.*, vol. 21, Nov. 1973.

[13] P. A. Bello, 'Characterization of Randomly Time-variant Linear Channels,' *IRE Trans. Commun. Syst.*, vol. CS-11, Dec. 1963.

[14] R. J. C. Bultitude and G. K. Bedal, 'Propagation Characteristics on Microcellular Urban Mobile Radio Channels at 910 MHz,' *IEEE J. Select. Areas Commun.*, vol. 5, Nov./Dec. 1987.

[15] D. M. J. Devasirvatham, R. R. Murray and D. R. Wolter, 'Time Delay Spread Measurements in Wireless Local Loop Test Bed,' *Proc. VTC'95*, pp. 241–245.

[16] K. L. Blackard, M. J. Feuerstein, T. S. Rappaport, S. Y. Seidel and H. H. Xia, 'Path Loss and Delay Spread Models as Functions of Antenna Height for Microcellular System Design,' *VTC'92*, pp. 333–337.

[17] P. Bartolomé, 'Temporal Dispersion Measurements for Radio Local Loop Application,' *VTC'95*, pp. 257–260.

[18] H. H. Xia, 'Simplified Model for Predicting Path Loss in Urban and Suburban Environments,' *IEEE Trans. Veh. Technol.*, vol. 46, no. 3, Nov. 1997.

[19] H. L. Bertoni, W. Honcharenko, L. R. Maciel and H. H. Xia, 'UHF Propagation Prediction for Wireless Personal Communications,' *Proc. IEEE*, vol. 82, pp. 1333–1359, 1994.

[20] J. B. Keller, 'Geometrical Theory of Diffraction,' *J. Opt. Soc. Amer.*, vol. 52, pp. 116–131, 1962.

[21] P. C. Gailey and R. A. Tell, 'An Engineering Assessment of the Potential Impact of Federal Radiation Protection Guidance on the AM, FM, and TV Broadcast services,' the U. S. Environmental Protection Agency (EPA), Report No. EPA 520/6–85–011, April 1985.

[22] D. M. Namra, C. W. I. Pistorius and J. A. G. Malherbee, *Introduction to the Uniform Geometrical Theory of Diffraction*, Artech House, London, 1990.

[23] H. L. Bertoni, in private communications.

[24] M. Abramowitz and I. A. Stegun, *Handbook of Mathematical Functions*, pp. 301–302, Dover, 1965.

[25] L. R. Maciel, H. L. Bertoni and H. H. Xia, 'Unified Approach to Prediction of Propagation Over Buildings for All Ranges of Base Station Antenna Height,' *IEEE Trans. Veh. Technol.*, vol. 42, pp. 41–45, Feb. 1993.

[26] J. Walfisch and H. L. Bertoni, *UHF/Microwave Propagation in Urban Environments*, Ph. D. Dissertation, Polytechnic University, 1986.

[27] H. L. Bertoni, *Radio Propagation for Modern Wireless Applications*, Prentice Hall PTR, in press.

[28] F. Ikegami, S. Yoshida, T. Takeuchi and M. Umehira, 'Propagation Factors Controlling Mean Field Strength on Urban Streets,' *IEEE Trans. Antennas Propagat.*, vol. 32, no. 8, pp. 822–829, Aug. 1984.

[29] Telesis Technologies Laboratory, 'Experimental license progress report,' to the Federal Communication Commission, pp. 5–70, Aug. 1991.

[30] H. H. Xia, H. L. Bertoni, L. R. Maciel, A. L. Stewart and R. Rowe, 'Microcellular Propagation Characteristics for Personal Communications in Urban and Suburban Environments,' *IEEE Trans. Veh. Technol.*, vol. 43, pp. 743–752, Aug. 1994.

[31] H. H. Xia, H. L. Bertoni, L. R. Maciel, A. L. Stewart and R. Rowe, 'Radio Propagation Characteristics for Line-Of-Sight Microcellular and Personal Communications,' *IEEE Trans. Antennas Propagat.*, vol. 40, pp. 170–177, Feb. 1992.

[32] T. A. Russel, C. W. Bostian and T. S. Rappaport, 'Predicting Microwave Diffraction in the Shadows of Buildings,' Virginia Polytechnic Inst. & State Univ., MPRG-TR-92-01, p. 40, Jan. 1992.

[33] J. Boersma, 'On Certain Multiple Integrals Occurring in Waveguide Scattering Problem,' *SIAM J. Math. Anal.*, vol. 9, no. 2, pp. 377–393, 1978.

3

Wireless Local Loop Networks Capacity Enhancement by Space Division Multiple Access

Giselle M. Galvan-Tejada and John G. Gardiner

3.1 Introduction

The provision of wireless access to the *Public Switched Telephone Network* (PSTN) from a customer premises is known as *Wireless Local Loop* (WLL). The potential markets of application of WLL range from developing countries to developed ones. In both cases, the liberalization of the telecommunications market has allowed competition among operators around the world. Many communities of difficult access are still waiting for basic telephone service, which could be viable only by wireless technology, which, besides, offers a fast and cost-effective option. However, as all wireless systems, WLL must face the task of employing efficiently the scarce radio frequency spectrum, within a services demanding market. This is translated into the need for a multiple access technique capable to provide a capacity as large as possible. *Space Division Multiple Access* (SDMA) is being widely considered as a mechanism to achieve this aim. SDMA is supported on the use of an adaptive antenna array at base stations. Based on SDMA principles, an antenna array at WLL customer premises is proposed here in order to get an additional improvement in the system capacity.

3.2 Background

3.2.1 Definition of Wireless Local Loop and General Aspects

3.2.1.1 Why Wireless in the Local Loop

The classical telephone network (Public Switch Telephone Network, PSTN) consists of user terminals and switching centres or exchanges. The user terminals are connected through several exchanges by means of wired links. Usually the end link (i.e. between the last exchange and the user) has been traditionally implemented with copper wire. This last link is known as the *local loop*. Thus, a wireless local loop will be one that uses the radio technology to access from a user terminal to its local exchange, as Figure 3.1

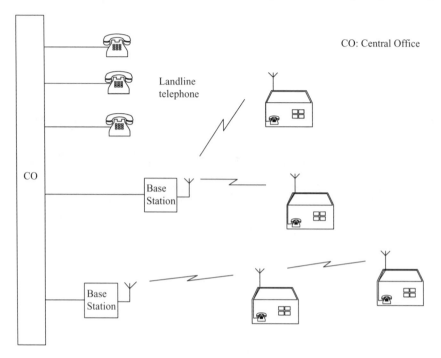

Figure 3.1   Representation of a WLL system introduced in a PSTN network

shows. For this reason, sometimes it is referred to as *Radio Local Loop* as well [1]. On the other hand, since the user terminal is fixed, the wireless local loop is also known as *Fixed Wireless Access* or *Fixed Radio Access* [1].

But, when and why did wireless local loop first emerge? The historical background of WLL is found in the early 1950s when terrestrial microwave links were developed to provide telephone access to users in rural areas. During the next four decades, different equipment, systems and technologies were tested in order to get a profitable solution and the idea of implementing wireless in the local loop began to consolidate as reality. Nevertheless, it was not until the reunification of Germany that the WLL concept became established [1]. The need for a fast and economic technological solution, along with the worldwide tendency of competition in the telecommunications market, boosted the growth of WLL systems.

Thus, the introduction of a new technology in the local loop may be justified for two different basic reasons. Firstly, a new technology may substitute old components of an existing network and to improve the ratio of cost to performance. Secondly, the technology may enable totally new services and applications to be implemented, offering competitive advantage to network operators. Both of these arguments are applicable to radio access.

Little regulatory work has been done to standardize WLL. In fact, WLL is currently in the so-called 'first generation' [1], so there are no defined standards around the world. In this matter, some modifications to mobile cellular systems were proposed by the CCIR [12]. In 1994 the *European Telecommunications Standards Institute* (ETSI) published the first report about WLL systems [4], where various aspects such as technological alternatives, operational characteristics, among others, are presented. In that report, studies of radio propagation for WLL conditions were recommended as a priority.

3.2.1.2 Propagation Issues

In any wireless communication system propagation studies are required in order to model the radio channel. The propagation characteristics will depend on the operation frequency and the environment where the system is working (for instance, mobile or fixed) mainly. The WLL system is no exception and some measurements have already been carried out in different conditions such as residential, business and industrial areas [21,22]. Although a WLL system is, in principle, able to accommodate certain mobility, its nature is primarily fixed. Hence, it is important to pay attention to those aspects related to radio propagation under static conditions.

For instance, by examining the path loss, fixed links (like terrestrial microwave systems) are planned in such a way that usually ground effects are avoided, even under difficult conditions. This is achieved by taking into account subrefractive situations and by a careful choice of antenna heights, ensuring that the height of the radio path over obstacles be allowed for a clear Fresnel zone. This approach leads to a fixed path loss between antennas, a situation that the operator cannot improve.

Another important aspect is the multipath fading phenomenon caused by delayed replicas of the transmitted signal that could introduce certain degradation in the system performance. A suitable choice of the geographical sites must be taken into consideration to avoid deep fades for long time periods that can happen in fixed situations like wireless local loop.

3.2.1.3 Services and Operational Characteristics for WLL

The service attributes and the operational characteristics for WLL systems are dealt with in [4] according to their particular characteristics, which, in part, are related to the type of service that it is expected they will provide.

Service attributes

- *Traffic Requirements*. Typical values of traffic are 70 mErl for residential lines and 150 mErl for business lines.
- *Access Network Delay*. This delay corresponds to that introduced by the radio circuits in the local loop. In spite of the absence of an established maximum value for this delay for WLL, a delay as short as possible is recommended to provide an acceptable voice service.
- *Grade of Service* (GoS). This figure represents the blocking probability of a system. The value recommended for WLL is 10^{-2} [4].
- *Lost Calls*. Under heavy traffic load (even exceeding the designed capacity), established calls should not be lost and blocking in the network should be in accordance with the specified GoS.
- *Service Security and Authentication*. As in any radio system, WLL should consider the implementation of some form of ciphering as a mechanism both to guarantee a secure communication and to authenticate the user into the network.
- *Mobility*. This attribute is only considered as a potential option in the future WLL systems, for which new considerations in terms of GoS will have to be taken into account.

- *Service Transparency.* Performance measurements in the communications link should be maintained as in conventional wired networks, e.g. a bit error rate, BER $< 10^{-3}$ for voice and a BER $< 10^{-6}$ for data. Other characteristics of standard wired networks, like number plan, network tones, just to mention a few, should also be transparent for WLL users.
- *Voice, Data and Multimedia.* First of all, WLL systems should be able to provide voice service with wireline quality or better. Fax, ISDN and Internet access at higher data rate are becoming important services today. The growing tendency for using multimedia impacts on the characteristics of a WLL system as well. Some considerations have recently been taken on the matter [18,19].

Operational characteristics

- *Frequency Efficiency.* Due to the limitations in available spectrum, diverse considerations should be taken into account in order to make efficient use of the allocated bandwidth. These include proper modulation formats, multiple access schemes, channel allocation plans, among others.
- *Radio Range.* WLL systems should be able to provide service to different user densities (urban, suburban and rural areas). Consequently, different ranges of coverage are expected, which, in turn, will be constrained by the equipment used. It is worth mentioning that relay operation should be considered as a mechanism to extend coverage in sparsely populated areas or in cases of low range equipment.
- *Radio Termination Characteristics.* ETSI establishes certain technical requirements for the radio termination at the customer premises. Among these, it is worth highlighting the power supply, the antenna mounting (external or internal antenna) and the capability of monitoring some general parameters of the system (like link quality, battery state, etc.). The power supply is very important in WLL because the operator cannot supply it from the base stations as happens in the wired network.
- *Radio Safety and Electromagnetic Compatibility* (EMC). All equipment employed should comply with the international standards of maximum permissible levels of exposure to electromagnetic fields [4]. With respect to the EMC considerations, WLL systems should meet the established protection levels in order to avoid disturbing (or being disturbed by) other already working systems as well as electrical and electronic equipment around.

3.2.2   Reference Model for a WLL System

The WLL reference model, which is independent of the technology applied, was defined by ETSI as Figure 3.2 shows ETSI ETR 139 [4]. In general terms, ETSI proposed that a WLL system might consist of the following elements and interfaces:

Local Exchange (LE)   In this model 'local exchange' is intended to represent a number of different elements of the PSTN network, according to operator requirements. These include the telephone network, leased line network and data network.

Base Station   One or more base stations may be connected to the controller. Each of them receives and transmits information and signalling from/to a customer terminal; they must also monitor the radio path.

Radio Termination   The radio termination has the ability to access the air-interface. It should be possible to support standard ISDN, PSTN or leased line terminals via the radio termination.

Customer Terminal   A standard ISDN or PSTN terminal.

Network Management Agent (NMA)   This element handles configuration data, customer, system and radio parameters.

Controller   The functions of this entity are to control the base stations, interface to the NMA element, and connect the WLL into LE/PSTN.

LE to Controller Interface, IF1   This interface connects the WLL access network to the public fixed network. The information carried by IF1 interface is related to the services accessed by the WLL users.

NMA Interface, IF2   Interfaces the NMA and the controller.

Controller to BS Interface, IF3   Connects one or more BSs to the controller; information related to the call handling, radio resource management, O&M messages, and mobility management specific for WLL.

Radio Interface, IF4   This interface carries the same information as the IF3 interface. In addition, it may be used to carry supervisory messages to the radio termination.

Radio Termination to Customer Terminal Interface, IF5   Information related to the services accessed by a user or an application is carried in this interface.

O&M Interfaces, IF6   Information related to the configuration, performance and fault management of the WLL system is carried in this interface.

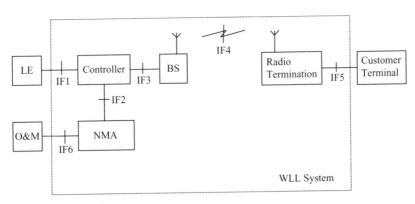

LE: Local Exchange
BS: Base Station
NMA: Network Management Agent
O&M: Operation & Maintenance Centre

Figure 3.2   Reference Model

3.2.3 WLL Deployment Examples

Some examples where a WLL system could be applied were treated in the early report published by ETSI [4]. These examples show several possibilities that interested operators could consider.

- *Existing operators in a new area.* Figures 3.3 and 3.4 depict two examples of implementation of a new wireless local loop service both for the case of a new housing area near to an existing network and the situation of a new town growing and hence requiring some telecommunications services.
- *Replacement of obsolete copper lines in rural areas.* In this case the main problem is that the communications services in rural areas have high maintenance cost and poor quality. For this reason, it is convenient to group those rural areas being relatively near each

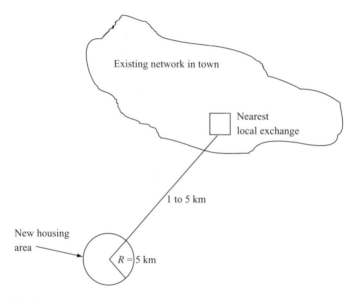

Figure 3.3 Example of new housing area with necessities of a near existing network

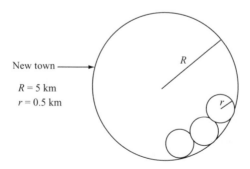

Figure 3.4 Example of a new town that requires of voice and data services

other, in one radio cluster whose central station will be able to attend them. It is worth noting that some places in the cluster could need other kinds of services, e.g. fax, data transmission, etc. and improve the old telephone network to provide a better voice quality.

- *Increase of capacity of an existing network.* This is the case when an existing cable network has reached its capacity limit and an easy expansion is not possible. If there are customers who require voice and low speed data services, they can change to WLL option leaving their copper lines for other users.
- *New operator in a competitive environment.* This situation is not very profitable at all, because the initial costs of rolling out a new network can be so high that in the early development stage just a few users are disposed to pay for such a service [2]. However, the potential market of users is very attractive for WLL operators. Thus, the main challenges that an operator should consider are:

— Requirement to cover customers in a variety of areas (urban, suburban, and rural areas).
— Needs to achieve coverage quickly over the maximum possible area.
— No existing infrastructure in place.
— The offered service must be technically equivalent to that provided by existing operator, probably using a wired network.

3.3 Technologies for the First Generation of Wireless Local Loop

Different technological approaches are actually contending to prove that they are the best choice for WLL. Cordless, cellular and proprietary technologies have been classified by W. Webb [1] as principal candidates to provide WLL facilities. In turn, several commercial systems with their own characteristics (multiple access technique, modulation, etc.) have adopted some of aforementioned technologies. Satellite-WLL has been also included in a classification given by A. R. Noerpel and Y. B. Lin [2] and it has been a topic of study in workshops like [3]. However, from a strict definition of WLL, satellite technology is just a mechanism to extend the coverage of a WLL network for rural areas. Therefore, only cordless, cellular and proprietary technologies are briefly considered next.

3.3.1 Cordless Systems

In their origin cordless systems were designed for indoor environments, so having a limited coverage and low power. This short range gives also the possibility of high data rates and simple implementations. In spite of their coverage constraint, they have gained interest as an alternative for WLL [4–10], and even propagation studies for outdoor environments have been carried out [11].

Certain technical issues of cordless systems could be modified in order to make them suitable for WLL systems. For instance, the receiver sensitivity specification could be significantly improved to extend the range for the WLL application; the transmitter output power could also be increased to provide greater range. Nevertheless, this option must be taken carefully since co-existence with other electronic and electrical equipment as well as communication systems must be free of disturbances; another alternative is that cordless systems have relay operation for extending radio range without the need for landline connections.

The *Digital European Cordless Telecommunications* (DECT) standard, the *Personal Access Communication System* (PACS) from USA, and the Japanese *Personal Handy-phone System* (PHS) are strong cordless candidates for WLL applications on high-density areas.

Dect [1,34] In 1992 the DECT standard was completed by ETSI. This standard was launched as an alternative to supersede the Cordless Telephony series, like CT-2. Operating in the band 1880–1900 MHz, the DECT system enables users to make and receive calls within range of base stations around 100 m in an indoor environment and up to 500 m in an outdoor environment. DECT uses a multicarrier TDMA/TDD format for radio communications between handset and base stations. This system provides 10 radio carriers of 1.728 MHz wide. Each of them is divided into 24 time slots, two of which provide a duplex speech channel. Each time slot runs at 32 kbps; each channel (including signalling and overhead) operates at a speed of 1.152 Mbps.

When a call is set up, 2 of the 24 time slots are used, alternating between transmitting and receiving signals. The remainder of the time can be used by the handset to monitor all other frequencies and time slots and transfer the call to a better speech channel if there is one available.

It is worth mentioning the channel selection feature of the DECT system. Channels need not be previously assigned to cell sites by making use of the *dynamic channel allocation* (DCA) scheme. Thus, the handset may activate any channel it determines to be free. In this way, DECT systems can quickly adapt to changes in propagation conditions or traffic load.

As far as field trials are concerned, since 1990 different operators have tested cordless equipment around the world in order to assess performance in a wireless local loop environment [33]. For instance, in 1994 a trial was developed in a small location of Norway. The main objective of the trail was to test the DECT ability to operate in a multioperating environment because in that location there were residential, down-town, business and industrial areas, all concentrated within a small location. The DECT trial system consisted of 160 radio base stations and some 240 handsets linked to the system. The average distance between a base station and a customer's premises was 80 m. This field trial was successful not only for simple fixed wireless access, but also for wireless access with mobility.

Another test cordless equipment was installed in the region of Aalborg University, Denmark [33], where there were both private and business customers within the expected range of a little more than 1 km. The radio base station antenna arrangement comprised three sectors each with 15 dBi directional antennas, supplemented by one 10 dBi omni-directional antenna for redundancy. Subscriber installations employed either 8 dBi directional or 2 dBi omni-directional antennas. The more important result is: Fax transmissions at data rates of 4800 bps were comparable in speed and quality to the wired network, however at 9600 bps the quality was degraded, hence the customer's modems were forced to work in a range between 4800 and 7200 bps. Nevertheless, in October 1995 a marketing study showed that a lot of customers were satisfied with these services.

PHS [1] As a solution to provide mobile service in very high-density pedestrian areas, the Research and Development Centre for Radio Systems of Japan developed the PHS standard, which was specified in 1993. Systems based on the PHS air-interface operate in 1895–1906.1 MHz band for home and office applications, and from 1906.1 MHz to 1918.1 MHz for public pedestrian environments (typically up to 500 m). In each of the 300 kHz wide RF carriers there are four traffic channels from which one is a dedicated control channel. A PHS system works on a TDMA/TDD base, whose frame duration is 5 ms; the modulation format is $\pi/4$-DQPSK at a channel rate of 384 kbps; a 32 kbps ADPCM voice encoder is utilized to provide wireline quality. In order to avoid frequency planning, this standard is specified to use DCA.

PACS [10] Based on the experience of the PHS proposal and the Wireless Access Communication System developed by BellSouth in the USA, in 1996 PACS emerged as one of the *Personal Communications Systems* (PCS) standards of the American National Standards Institute in USA. This standard was created to provide a short-range service (typically 500 m) in both WLL and mobile cellular environments. There are two frequency bands in which PACS operates: the licensed PCS band, allocated in 1850–1910 MHz for the uplink and between 1930 and 1960 MHz for the downlink; and the unlicensed PCS band in the range from 1920 MHz to 1930 MHz. The channel spacing is 300 kHz for both bands. PACS systems operate in FDD and TDD modes for the licensed PCS band and the unlicensed PCS band respectively.

The user stations can be portable *subscriber units* (SUs) or *wireless access fixed units* (WAFUs) which have access to the *radio ports* (RPs) by means of TDMA. The duration of the PACS frame is 2.5 ms which is divided in 8 time slots. In the FDD mode 7 time slots are used for voice/data transmissions and one for control. In the unlicensed band case the 8 time slots are used as four two-way channels since the control channel is allowed to be used for voice/data transmissions when all channels are busy. The transmitted signals are modulated in $\pi/4$-DQPSK format; PACS provides wireline quality by employing 32 kb/s ADPCM as voice encoder.

The frequency allocation scheme is known as *Quasi-Static Autonomous Frequency Assignment* (QSAFA), which eliminates the need for precise frequency planning. A controller is installed at each RP. So, each RP autonomously measures the signal strength, and updates and selects the best channel frequency for a given signal-to-interference threshold. This scheme avoids the need of a central control.

A study of traffic and coverage in two cities of Florida, USA was carried out to assess the initial deployment cost for a WLL-PACS network [7]. In that study, the total number of RPs required as a function of the WLL penetration was taken as parameter of assessment.

The first case was Miami, which represents a large area with high population and traffic density. In that case, the number of RPs required grows steadily as the WLL penetration does. The second case was the city of Jacksonville, where there is a medium population and traffic requirement. In that case, a slower incrementation of the number of RPs required was observed.

3.3.2 Cellular Technologies

Cellular radio mobile systems are ever more popular around the world not only to provide mobile service, but also as a rapid solution to provision of telephone service in areas of difficult access [13,14]. In general terms, cellular systems seem to be a natural option for

WLL due to their infrastructure already developed and the large coverage that they possess. So, extensive considerations have been taken to make the proper modifications for WLL and several performance analyses have been studied [4,12,13,14,16]. The disadvantages of cellular technologies applied for WLL are, firstly, the poor voice quality, and then low data rate, complexity and cost. However, cellular-WLL operators claim that it is worth adjusting to WLL necessities in order to make efficient use of cellular infrastructure [16].

A WLL service could be added to a mobile network as a specific service profile with restricted or denied mobility, modified numbering plan, special charging, etc. In the development of the WLL system, the possibility of supporting different access types (either by mobile subscriber, wired subscriber or WLL subscriber) in the same access network could be included and both Germany and Spain have used analogue cellular systems to share mobile and WLL users [4]. However, the main objection or difficulty in this combined implementation is the impossibility of satisfying high data rate requirements.

To date, the most representative cellular systems are the European *Global System for Mobile Communications* (GSM), and the *Interim Standard 54* (IS-54) and IS-95 standards from USA. A recompilation of their most representative characteristics as well as studies of possible implementations of cellular-WLL networks is presented next.

GSM and DCS-1800 [1,34]   The digital cellular system adopted in Europe is the Global System for Mobile Communications (GSM) whose network began to work in 1991. This system operates in a TDMA mode and uses a *regular pulse excited linear predictive coder* (RPE-LPC) to encode the speech at 13 kbps. The data are transmitted via GMSK at a TDMA rate of 270.8 kbps. The frequency bands are 935–960 and 890–915 MHz. The carrier spacing is 200 kHz.

The ETSI GSM technical committee developed a modified version of GSM, known as *Digital Cellular System at 1800 MHz* (DCS1800). Some variations of this system include duplex bands of 75 MHz with 20 MHz guard band and smaller cells than those used in GSM, which implies lower power levels.

The GSM standard describes all necessary aspects of a digital cellular system. Therefore, this standard could be used in various ways for WLL. There are three ways to implement a WLL service with GSM infrastructure [4]: (1) Use of the GSM/DSC 1800 network for WLL; (2) use of a localised GSM network for WLL; (3) use of the GSM air interface for WLL.

For example, a proposal for WLL based on the GSM/DCS air-interface is presented in [16]. Several environments (urban, suburban and rural areas) were simulated according to measurements carried out in Finland. Some of the characteristics of that system include carrier frequency of 1850 MHz, use of antenna diversity and frequency hopping. By combining slow frequency hopping and antenna diversity, a good performance of the WLL system in every simulated environment was obtained; other results include isolated cases, e.g. the use only of slow frequency hopping, which offered a quite good performance for the urban and suburban cases, but not for rural environment.

IS-54 [34]   In the United States, the analogue *Advanced Mobile Phone System* (AMPS) was modified to evolve towards a digital system known as D-AMPS (standardized as IS-54). The sudden boost of mobile users and the high cost of the cell sites created the need of a digital solution. Thus, the IS-54 standard fits three TDMA 8-kbps encoded speech channels into each 30 kHz AMPS channel.

IS-54 uses a linear modulation technique DQPSK to provide good bandwidth efficiency. The transmission rate is 48.6 kbps with a channel spacing of 30 kHz. This gives a bandwidth efficiency of 1.62 b/s/Hz, The main penalty of linear modulation is power efficiency that affects the weight of handsets and time of battery charging. IS-54 uses a VSELP speech coder, chosen for its fast operation, high quality voice, robustness to channel errors and modest complexity [34]. The source rate is 7.95 kbps and the transmission rate is 13 kbps.

Qualcomm CDMA and IS-95 [34]   Qualcomm developed the IS-95 CDMA system as an alternative to the TDMA cellular systems. The access technique CDMA is based in the spread spectrum property. This consists of assigning one code to each user terminal signal. When this is done, the power spectrum is spread and for this reason this process is known as spread spectrum. The CDMA codes are generated by Walsh functions that are mathematically shown to form an orthogonal set. Thus, any couple of transmitters using different codes may be identified by means of a correlation process. The spread spectrum signals have many advantages that are quite attractive from several perspectives. For example, these signals are effective in mitigating multipath fading because their wide bandwidth effectively introduces frequency diversity.

Qualcomm CDMA operates between 824 and 849 MHz for reverse link and between 869 and 894 MHz for the forward link. The CDMA bandwidth required for each up- and down-link is 1.23 MHz. This CDMA network also operates in the 1.7–1.8 GHz band. Since CDMA channels can be reused in adjacent cells, no frequency planning is required. The Qualcomm proposal is divided into two parts: the *core* and *extended CDMA systems*. The core system uses a 14.4 kbps air-interface and 13 kbps codecs (voice coders). The extended CDMA system uses a 76.8 kbps air-interface and works with higher quality codecs at 16 and 32 kbps.

In each up- and down-link there are 64 channels, which are coded by Walsh functions. Each symbol generates an appropriate 64-chip Walsh code which is then combined with a pseudorandom bit sequence to bring the rate up to 1.2288 Mchip/s. All the 64 CDMA channels are combined to give a single I and Q channel. These signals are applied to quadrature modulators and the resulting signals summed to form a CDMA/QPSK signal.

3.3.3   Proprietary Technologies

The lack of a WLL standard has allowed the emergence of independent technologies called 'proprietary'. The principle of these approaches is the development of a technology exclusively designed for WLL necessities, instead of adapting cellular or cordless systems. Thus, these technologies give emphasis to the quality of service required for WLL [4]. They have the flexibility of selecting their own technical parameters (modulation format, multiple access scheme, etc.) limited by spectrum and interference restrictions, and they do not need to cover un-populated zones (operators call this characteristic 'coverage by islands' [17]). Nevertheless, all proprietary technologies are yet in an early stage and their impact cannot be easily assessed. Among principal proprietary technologies are found: *Nortel Ionica, Tadiran Multigain, DSC Airspan* and *Lucent Airloop* [1].

Nortel's Ionica   Motivated by the liberalization of the telecommunications market in the UK, Ionica (known around the world as Proximity I) was developed in the early 1990s and commercially launched in 1996 making it one of the first proprietary WLL system [1]. The Ionica WLL system is based on the benefits of using a technology specifically designed for WLL instead of trying to adapt a cellular mobile or cordless system to the needs of WLL. The features of the Ionica system are listed below in Table 3.1 [17]:

As with any proprietary technology, Ionica takes advantage of the flexibility in the choice of its parameters of design. For instance, the use of a directional antenna at the customer's RSS makes a substantial difference relative to mobile cellular or cordless systems. The high gain given by directional antennas allows greater coverage (typical range of 15 km [1], although up to 35 km in line-of-sight paths has been claimed [17]. This type of antenna provides also a considerable reduction of the effects of the multipath fading. However, due to fact that multipath can never be entirely eliminated, the Ionica system has adopted the use of an equalizer [17].

Table 3.1   Ionica system specifications

Parameter		Comment
Frequency bands	3425–3442 MHz	*Residential Service System* (RSS) transmitter
	3475–3492 MHz	*Base Station* (BS) transmitter
RF channels	54	
Channel spacing	307.2 kHz	
Access/Duplexing Technique	TDMA/FDD	
TDMA structure	10 slots/frame	
Frame duration	5 ms	
Slot duration	500 μs	
Modulation	π/4-DQPSK	RRC filtering, $\alpha = 0.4$
Net user bit rate/time slot	32 kbps	user can access up to 3 time slots/frame
BS transmission power	+29 dBm	measured at the antenna connector
BS sensitivity	−102.5 dBm	at the antenna connector, for BER $< 10^{-3}$ Pre FEC
RSS EIRP	+45 dBm	integral antenna gain = 18 dBi
RSS sensitivity	−102.5 dBm	at the antenna connector, for BER $< 10^{-3}$ Pre FEC
RSS antenna beamwidth	20°	typical, at 3 dB
Speech coding	32 kbps ADPCM	ITU-T G.721

DSC Airspan Considered as the first commercially available WLL system, Airspan (engineered by Airspan Communications Corporation, formerly DSC) is a spread spectrum technology developed to provide service in rural areas at 2 GHz [1]. In order to show the potential of Airspan, a small system of five subscribers took part in trials conducted by *British Telecomm* (BT) in 1993. The success of those trials led to the whole development of the commercial product in 1995.

This technology operates in CDMA/FDD mode. The carrier spacing is 3.5 MHz, within which up to 15 channels of 160 kbps are provided [1]. Initially, Airspan provided voice service using 64 kbps PCM, but more recently a voice encoder ADPCM at 32 kbps has been adopted [1]. The maximum transmission power of the base stations is 33 dBm. At the moment, some parameters of this system, like the modulation format and the chip rate, are not available in the open literature.

Lucent's Airloop Airloop is a CDMA technology developed in the UK as a competitive option to its counterpart Airspan. This technology has been planned to operate in the 3.4 GHz band (other frequency bands have been considered around the World, such as 1900 MHz PCS band in the USA and 2.3 GHz bands in Europe) with channels 5 MHz wide. Within this bandwidth there are 115 channels at a bit rate of 16 kbps. The spreading factor is 256 giving a chip rate of 4.096 Mcps. 32 kbps ADPCM has been chosen as the voice encoder [1]. The modulation format is QPSK. The transmission power of the base stations is 35 dBm and the coverage range varies from 2.5 km to 6 km depending on the environment.

As far as works carried out for Airloop are concerned, a time-controlled DS-CDMA design is proposed in [15]. This proposal takes advantage of the fixed environment of a WLL system to achieve an accurate synchronization and closed loop time control of the reverse link transmissions. Thus, a degree of suppression of the multiple access interference experienced on the reverse link is accomplished, increasing the number of subscribers that can be served with high quality of service.

All the above technologies adopted for WLL systems are included in the so-called 'first generation of WLL' [1], and it is expected that a second generation of WLL contemplates multimedia services. In effect, although voice is the primary service that a customer requires, recently a growing interest in employing multimedia services in WLL has appeared as well [18,19]. These services need high transmission rates and consequently wider bandwidth occupancy. If the signal bandwidth is higher than the coherence bandwidth of the channel (which has an inverse relationship with the delay spread [20]), the channel will exhibit a frequency selective behaviour, causing intersymbol interference [20]. The fixed nature of WLL systems eases the implementation of directional antennas at *user station*s (USs), thus providing a beneficial alternative to avoid a large delay spread due to multipath [21,22]. Furthermore, a base station (BS) can use an adaptive antenna array in order to form directional beams towards USs. Thanks to this, simpler systems can be implemented, possibly avoiding the use of complex equalization techniques for instance.

Here, a modification to the user station antenna is proposed, keeping the above advantages of a directional antenna (as is commonly done in the practice, e.g. [17,23]), but giving to the user the potentiality of selecting among several base stations. This is done with an antenna array, which is the fundamental idea in this chapter. In order to contrast this proposed configuration, other structures are analysed and compared (including the conventional one employed in WLL) as will be seen later.

3.4 Multiple Access Techniques

Traditional multiaccess schemes are *Frequency Division Multiple Access* (FDMA), *Time Division Multiple Access* (TDMA) and *Code Division Multiple Access* (CDMA). Recently, *Space Division Multiple Access* (SDMA) has emerged as a promising possibility to use the radio spectrum efficiently.

In the context of WLL systems, operators have adopted different multiaccess schemes, from analogue FDMA to digital TDMA and CDMA. Those WLL systems based on cellular or cordless standards apply the multiple access mechanism appropriate to the chosen standard. On the other hand, operators offering WLL service based on proprietary technology can select their access technique according to their own considerations. In any case, a multiple access scheme dictates the system capacity and the use of the scarce spectrum, which is of major concern for the operators.

3.4.1 FDMA

One of the main problems in a wireless system is how to use the available spectrum to provide a given service. The simplest technique is to divide it in a certain number of frequency bands, which is the principle of FDMA. In general, these frequency bands represent the traffic channels and in some cases, control channels. Each of these channels is used during all the call. As a consequence, one transceiver is dedicated per channel and even if there is no user requesting a channel, all transceivers have to be turned on all the time (which implies a waste). On the other hand, however, a system operating with FDMA does not need to be synchronized.

This technique is constrained to have good quality RF filters in order to avoid overlapping between adjacent channels. Due to the fact that it is not possible to get a perfect partition between adjacent channels, it is common to distribute them in non-adjacent cells to reduce the effects of adjacent channel interference.

3.4.2 TDMA

The arrival of digital technology allowed the possibility to use the radio spectrum in a different manner. Thus, instead of allocating each user a portion of the available spectrum during all the call, the time is divided in slots where one user per time slot is allowed to have access to the central station. In the basic form of operation of TDMA, the users can use the complete allocated spectrum but only in their corresponding time slot. Consequently, synchronization is vital for this technology; otherwise there could be non-desired collision among users.

3.4.3 CDMA

In this multiaccess technique, users can access the common server employing the whole available frequency band and during all the time (see Figure 3.3). The above is achieved by spreading the users' signals by multiplying each of them by different special sequences.

In other words, each user has a unique code with which it is identified. The special sequences or spreading codes have certain properties, which give to the receptor the possibility to distinguish a particular user in a multiuser environment. A big limitation in CDMA systems is that users must be power controlled to avoid the so-called *near–far* problem [35].

3.4.4 Space Division Multiple Access (SDMA)

Space division multiple access is a mechanism to make efficient use of the scarce radio frequency spectrum taking into account information of users' position. It provides a different domain than usual FDMA, TDMA, or CDMA techniques, because any resource that is being used (either frequency bands, time slots, codes, or any combination of them) can be theoretically shared by two (or more) users at a time in the same cell if and only if users have enough spatial separation and maintain an acceptable carrier-to-interference ratio [24]. Thus, SDMA is based on the application of an adaptive array antenna at the base station site, which is used to form multiple independent beams per traffic channel. As a result of narrow beams formed by antenna arrays, an extra benefit is accomplished: larger coverage range, which can be interchanged for a reduction in power consumption.

From all aforementioned advantages, the implementation of SDMA in WLL base stations has been taken already into consideration (see for example [25,26]). Moreover, experimental studies carried out by H-P. Lin et al. [27] show the stability of the array response in fixed environments which strongly benefits its application in WLL systems.

3.4.5 Digital Beamforming

In order to achieve the spatial separation operation it is fundamental that servers (base stations, etc.) have an antenna array and a digital signal processing stage (both devices together make up the so-called *digital beamformer network*, DBF [24]). As Litva states, the term beamforming describes the operation of a network to concentrate energy radiated towards a specific direction [24]. For this reason, sometimes these types of antennas are referred as adaptive or smart antennas.

Then, a DBF network separates the desired signal arriving at the terminal antenna array and suppressing the interfering signals. In principle, each user is identified in accordance with its geographical position (by means of the so-called *spatial signatures* same traffic channel. It is worth noting that 'enough spatial separation' does not necessarily mean geographically spaced users, but more precisely implies independent or *steering vector*), from which, and under certain criterion, some users can share the spatial signatures as they are seen by the antenna array. The steering vector represents the response of the antenna array to the wavefront impinging from a given direction on the elements of the array [36]. Moreover, there are other factors that influence the spatial separation capability. Broadly speaking, it is important to consider modulation format, carrier-to-interference ratios, etc., besides spatial signatures as factors that influence the allocation of a resource. In any case, a criterion is established so that its fulfilment guarantees the desired link quality.

3.4.6   Study scenarios

Case (1) Conventional Configuration   This basic configuration (widely employed by most WLL operators) exploits the fixed character of user stations by using directional antennas steered towards a certain base station [17,23]. The traditional 120° directional or omnidirectional antenna is used at base stations (Figure 3.5). Although directional antennas at customer stations improve the budget link and reduce the delay spread, under hard traffic load conditions, there is no mechanism by which another cell around the user can provide service to it, as occurs in mobile systems, where mobile stations use omni-directional antennas.

Case (2) Adaptive Array at Base Station Configuration   In the same form than the previous configuration, customer stations employ directional pre-planned antennas (i.e. in a fixed planning base), but now the base station has an adaptive antenna array. The provision of an adaptive antenna array at base stations has been widely considered as a mechanism to reduce cochannel interference in classical cellular planning [28]. In turn, an adaptive antenna array gives the possibility to work in the space domain by means of SDMA. The formation of high-gain individual beams steered towards each user from base stations plus the fact that user stations have directional antennas, reduces considerably the delay spread [29]. Nevertheless, the employment of directional antennas at user stations limits the selection of base stations by users, implying high blocking rates (even with dynamic channel allocation the situation does not improve for uniform traffic distribution [30]). For example, let us assume that in a WLL network each base station has up to 4 traffic channels to serve a call (including spatial processing). In Figure 3.6 a possible configuration is depicted. When user 5 tries to access base station 1 (which has

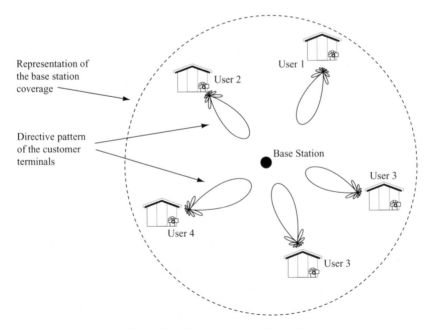

Figure 3.5   Conventional configuration

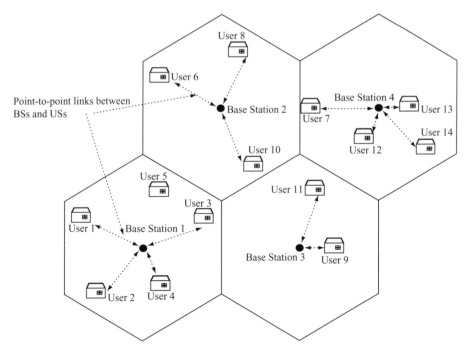

Figure 3.6 Adaptive antenna array at base station configuration

been previously allocated to attend that user), and that base station has all its channels busy, the call is blocked even though base stations 2 or 3 could connect with it.

Case (3) Omnidirectional Antennas at User Premises This scheme considers that the base station has an adaptive antenna array as in case 2, but an omni-directional antenna is assumed at subscriber station. This configuration offers to the user the possibility of choosing among several base stations, such that if a base station has reached its maximum capacity, a user can be served by neighbouring base stations. The above is illustrated in Figure 3.7, where user 5 would be now attended by base station 2. This is the configuration used by cordless systems like DECT (although usually DECT base stations employ traditional antennas), whose standard has been considered for WLL systems [31]. However, this configuration suffers from less coverage with respect to case 2, and large delay spreads due to omnidirectional reception at user stations.

Case (4) Proposed Configuration: Antenna Array at User Station An adaptive antenna array at base stations is again assumed in this configuration, but now it is proposed the employment of an antenna array at customer stations. This configuration gives flexibility to the user station to choose from base station candidates around, which are able to attend it under certain criterion. We propose that the user station scans and finds a proper base station by means of forming a narrow rotating beam. Once a given base station confirms to the user station the availability of a resource to attend its call, a point-to-point link is formed between the stations (see Figure 3.8), so reducing the delay spread compared to the one in case 3. Another advantage is the increase in coverage obtained from high-gain narrow beams in both links.

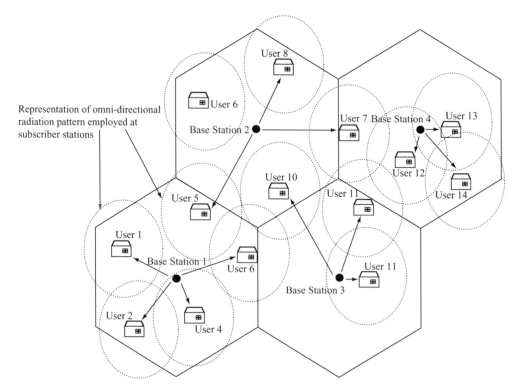

Figure 3.7   Omni-directional antennas employed at subscriber station

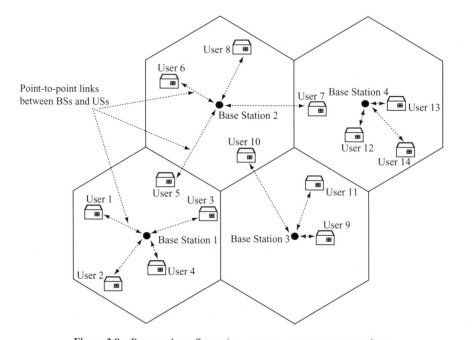

Figure 3.8   Proposed configuration: antenna array at user stations

The fixed user station has the advantage of making use of simpler technology for the antenna array than that of the base station, because it is not necessary for it to be an adaptive antenna array, which has to modify its pattern when a new user requires a call.

3.5 Simulation Description

The performance in terms of blocking probability for the four study cases presented in the previous section will be obtained. In order to do a fair comparison, equal conditions are taken in all cases.

First of all, one cluster of 4 cells was chosen whose base stations, localized at the centres of their corresponding cells, are provided with 5 resources each (frequency bands, for instance). A *fixed channel allocation* (FCA) planning without channel borrowing schemes was employed. The traffic model is based on Poisson-distributed call arrivals with holding time exponentially distributed (mean duration time of 120 seconds). Users were uniformly generated in the whole cluster, which was assumed to be square. There were no queues in the system, that is, blocked calls cleared.

Effects of shadowing and multipath fading were not taken into account. Thus, guaranteed links were contemplated in all cases, and, for cases 3 and 4 only, they were limited by the range given by the antennas at the user station. In other words, given that in cases 3 and 4 the users can choose more than one base station, depending on their positions, they will be able to access certain BS within a radius limited by their antenna's gain. Naturally, users with antenna arrays (and consequently greater gain than users with omni-directional antennas) have more opportunity to be connected with distant base stations.

A relationship between coverage distances for these two cases was found from the well-known Friis propagation expression, with a path loss exponent of 2. Let G_{omni} and G_{arr} denote the omni-directional antenna gain and antenna array gain (both in linear units), respectively. Let d_{omni} be the distance that the omni-directional antenna can cover ($d_{omni} = \sqrt{2}/2$ for the used structure, where base stations are distributed in the corners of a square with sides of unitary length) and let d_{arr} be the distance that the antenna array covers. By assuming in both cases the same reception and transmission power, as well as the same operating frequency, the relation between d_{arr} and d_{omni} as function of antenna's gain is given by

$$d_{arr} = d_{omni}\sqrt{\frac{G_{arr}}{G_{omni}}} \tag{3.1}$$

This expression is useful to model cases 3 and 4, because depending on the antenna's gain, a distinction of coverage is obtained. In our study, we set $G_{omni} = 3.98$ (6 dB) and $G_{arr} = 15.84$ (12 dB).

Finally, an adaptive antenna array able to separate 2 users in the same cell at the same time (cases 2, 3 and 4) was considered. In this concern, a minimum angular separation (α_{min}) was taken as the parameter to determine reusability of channels. Under this criterion, if two users have an angular distance greater than α_{min}, they can access the same channel by spatial division. Here, the spatial allocation scheme adopted is the named '*First Duplicate*' [32].

3.6 Performance Results

In order to validate our simulations, two theoretical models were taken into account. Erlang-B equation was used to contrast the simulation results in case 1. The analytic model developed in [32] served to verify the simulations in case 2. Both cases are shown in Figure 3.5, where P_s is the probability of successful spatial allocation [32], which is related with parameter α_{min} (here, $\alpha_{min} = \pi/8$ was chosen). The relation between P_s and α_{min} in this case is given by

$$P_s = \frac{\pi - \alpha_{min}}{\pi}, \quad \alpha_{min} \text{ in radians}$$

In this particular event, 10^5 calls were simulated.

As Figure 3.9 shows, results fit well with theoretical plots, validating our simulations. The plot labelled as 'Analytic' represents an increase of almost twice on capacity of the system [32]. Note that case 1 has a poorer performance than case 2 even though links are guaranteed. This implies more resources per cell for a grade of service (blocking probability) of 10^{-2} required for WLL systems [4].

In Figure 3.10 the performance of all cases is shown. Here, 10^6 calls were simulated. As was expected, the case 1 has the worst performance due to traditional antennas considered at base stations.

An interesting behaviour is observed in cases 2 and 3, whose performance plots are practically comparable. From these cases, a slight enhancement can be seen when user stations have omni-directional antennas (case 3), instead of directional (case 2). However, the advantage in terms of blocking probability accomplished by omni-directional antennas cannot be justified compared to the disadvantage of large delay spreads.

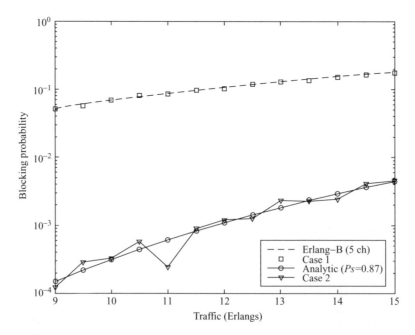

Figure 3.9 Comparison between theoretical models and simulation results

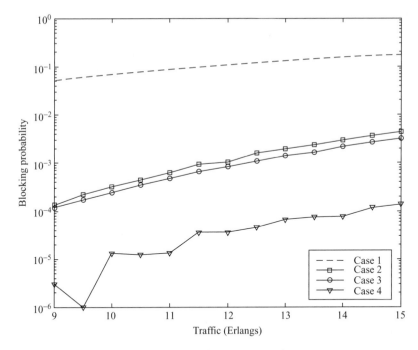

Figure 3.10 Comparison among study cases

Instead, as can be seen, our proposed configuration (case 4) reveals its superiority over the alternative configurations (three orders of magnitude with respect to the conventional case and almost two orders of magnitude in comparison with cases 2 and 3).

Although these results cannot be taken in an absolute base (due to simplifying assumptions), all of them were analysed and simulated under the same conditions, so that comparative performances are indeed valid.

References

[1] W. Webb, *Introduction to Wireless Local Loop*, Artech House, London, UK, 1998.

[2] A. R. Noerpel and Y. B. Lin, 'Wireless Local Loop: Architecture, Technologies and Services,' *IEEE Personal Commun.*, vol. 5, no. 3, pp. 74–80, June 1998.

[3] J. J. Spilker, I. W. Kane and M. C. Mertsching, 'Integrated VSAT OCDMA Wireless Local Loop for Rural Telephony,' *Asian VSAT Workshop*, Bankok, Thailand, 1996.

[4] ETSI ETR 139, 'Radio Local Loop,' Sophia Antipolis, Valbonne, France, Nov. 1994.

[5] H. M. Sandler, 'CT2 Radio Technology for Low Power Fixed Wireless Access,' in *IEEE International Symposium on Personal, Indoor and Mobile Radio Communications*, vol. 3, pp. 1133–1138, 1995.

[6] C. C. Yu, D. Morton, C. Stumpf, R. G. White and J. E. Wilkes, 'Low-Tier Wireless Local Loop Radio Systems—Part 1: Introduction,' *IEEE Commun. Mag.*, vol. 35, no. 3, pp. 84–92, Mar. 1997.

[7] C. C. Yu, D. Morton, C. Stumpf, R. G. White and J. E. Wilkes, 'Low-Tier Wireless Local Loop Radio Systems—Part 2: Comparison of Systems,' *IEEE Commun. Mag.*, vol. 35, no. 3, pp. 94–98, Mar. 1997.

[8] S. Kandiyoor, P. Van de Berg and S. Blomstergren, 'DECT: Meeting Needs and Creating Opportunities for Public Network Operators,' in *IEEE International Conference on Personal Wireless Communications*, pp. 28–32, 1996.

[9] M. Lotter and P. Van Rooyen, 'CDMA and DECT: Alternative Wireless Local Loop Technologies for Developing Countries,' in *IEEE International Symposium on Personal, Indoor and Mobile Radio Communications*, Helsinki, Finland, pp. 169–173, 1997.

[10] C. R. Baugh, E. Laborde, V. Pandey and V. Varma, 'Personal Access Communications System: Fixed Wireless Local Loop and Mobile Configurations and Services,' *Proc. IEEE*, vol. 86, no. 7, pp. 1498–1506, 1998.

[11] L. B. Lopes and M. R. Heath, 'The Performance of DECT in the Outdoor 1.8 GHz Radio Channel,' in *IEE 6th International Conference on Mobile Radio and Personal Communications*, 1991, London, UK, pp. 300–307, 9–11 Dec.

[12] CCIR Rec. 757, Basic System Requirements and Performance Objectives for Cellular Type Mobile Systems used as Fixed Systems, 1992.

[13] R. Westerveld and R. Prasad 'Rural Communications in India Using Fixed Cellular Radio Systems,' *IEEE Commun. Mag.*, vol. 32, no. 10, pp. 70–77, Oct. 1994.

[14] D. Anvekar, P. Agrawal and T. Patel, 'Fixed Cellular Rural Networks in Developing Countries: A Performance Evaluation,' in *IEEE International Conference on Personal Wireless Communications*, pp. 33–38, 1996.

[15] Q. Bi and D. R. Pulley, 'The Performance of DS-CDMA for Wireless Local Loop,' in *IEEE International Symposium on Spread Spectrum Techniques and Applications*, 1996, Mainz, Germany, pp. 1330–1333, Sep. 22–25.

[16] T. Westman, K. Rikkinen, T. Ojanpera and M. Tarkiainen, 'Wireless Local Loop (WLL) Based on DCS1800 Technology,' in *IEE Colloquium on Local Loop Fixed Radio Access*, pp. 3/1–3/6, 1995.

[17] R. McArthur, 'Ionica Fixed Radio Access System,' in *IEE Colloquium on Local Loop Fixed Radio Access*, pp. 5/1–5/11, 1995.

[18] J. Haine, 'HIPERACCESS: An Access System for the Information Age,' *Electronics Commun. Engineering J.*, vol. 10, no. 5, pp. 229–235, Oct. 1998.

[19] 'ATM Radio-in-the-Local-Loop (RL/A),' in *Proceedings of the Wireless Broadband Communications Workshop*, 1997, Brussels, Belgium, Sep. 29.

[20] W. C. Jakes, (Editor), *Microwave Mobile Communications*, IEEE Press, New York, 1974.

[21] Bartolomé, P., 'Temporal Dispersion Measurements for Radio Local Loop Applications,' *IEEE Vehicular Technology Conference*, vol. 1, pp. 257–260, 1995.

[22] I. J. Wassell, 'Delay Spread for a DECT Based Wireless Local Loop,' *SPIE The International Society for Optical Engineering Proceedings*, vol. 2601, pp. 255–262, 1995.

[23] F. G. Harrison, 'Microwave Radio in the British Telecom Access Network,' *British Telecommunications Engineering*, vol. 8, pp. 100–106, 1989.

[24] J. Litva and T. K. Y. Lo, *Digital Beamforming in Wireless Communications*, Artech House, London, UK, 1996.

[25] M. A. Abumahlula, T. L. Doumi and J. G. Gardiner, 'Adaptive Antennas and Mixed Traffic in a DECT-WLL Environment,' in *IEE Colloquium on Advanced TDMA Techniques and Applications*, London, UK, p. 60, 1996.

[26] B. Erickson, 'Adaptive Antenna Technology Aids Wireless Transmission,' *Microwaves & RF*, pp. 139–144, Dec. 1997.

[27] H. P. Lin, S. S. Jeng, I. Parra, G. Xu, W. J. Vogel and G. W. Torrence, 'Experimental Studies of SDMA Schemes for Wireless Communications,' in *IEEE International Conference on Acoustics, Speech and Signal Processing*, vol. 3, pp. 1760–1763, 1995.

[28] S. C. Swales, M. A. Beach, D. J. Edwards and J. P. McGeehan, 'The Performance Enhancement of Multibeam Adaptive Base-Station Antennas for Cellular Land Mobile Radio Systems,' *IEEE Transactions on Vehicular Technology*, vol. 39, no. 1, pp. 56–67, Feb. 1990.

[29] A. G. Burr, 'A Spatial Channel Model to Evaluate the Influence of Directional Antennas in Broadband Radio System,' *IEE Colloquium on Broadband Digital Radio: Challenge of the Radio Environment*, pp. 8/1–8/8, 1998.

[30] J. C. I. Chuang, 'Performance Issues and Algorithms for Dynamic Channel Assignment,' *IEEE J. Selected Areas Commun.*, vol. 11, no. 6, pp. 955–963, Aug. 1993.

[31] D. Akerberg, F. Brouwer, P. H. G. Van de Berg and J. Jager, 'DECT Technology for Radio in the Local Loop,' in *IEEE Vehicular Technology Conference*, vol. 2, pp. 1069–1073, 1994.

[32] G. M. Galvan-Tejada and J. G. Gardiner, 'Theoretical Blocking Probability for SDMA,' *IEE Proceedings-Commun.*, vol. 146, no. 5, pp. 303–306, Oct. 1999.

[33] W. H. W. Tuttlebee (Editor), *Cordless Telecommunications Worldwide: the Evolution of Unlicensed PCS*, Springer, London, 1996.

[34] T. S. Rappaport, *Wireless Communications*, Prentice Hall, New Jersey, 1996.

[35] J. C. Liberti and T. S. Rappaport, 'Analytical Results for Capacity Improvements in CDMA,' *IEEE Trans Vehicular Technology*, vol. 43, no. 3, pp. 680–690, Aug. 1994.

[36] S. Anderson, M. Millnert, M. Viberg and B. Wahlberg, 'An Adaptive Array for Mobile Communication System,' *IEEE Trans Vehicular Technology*, vol. 40, no. 1, pp. 230–236, Feb. 1991.

4

Combined Trellis Coded Quantization/Modulation over a Wireless Local Loop Environment

O. N. Uçan, M. Uysal and S. Paker

4.1 Introduction

In this chapter, combined trellis coded quantization/modulation scheme is introduced for wireless local loop environment modelled with realizable and practical medium parameters. The performance analysis of the combined system is carried out through the evaluation of *signal-to-quantization noise ratio* (SQNR) versus *signal-to-noise ratio* (SNR) curves and bit error probability upper bounds. Simulation studies confirm the analytical results.

4.2 Fundamentals of Trellis Coded Modulation

There is a growing need for reliable transmission of high quality voice and digital data for wireless communication systems. These systems, which will be part of an emerging all-digital network, are both power and band limited. To satisfy the bandwidth limitation, one can employ bandwidth efficient modulation techniques such as those that have been developed over the past several years for microwave communication systems. Examples of these are *multiple phase-shift keying* (MPSK), *quadrature amplitude modulation* (QAM), and varius forms of *continous phase frequency modulation* (CPM).

In the past, coding and modulation were treated as separate operations with regard to overall system design. In particular, most earlier works on coded digital communication systems are independently optimized: (1) conventional (block or convolutional) coding with maximized minimum Hamming distance (2) conventional modulation with maximally separated signals.

In a bandwidth limited environment, higher-order modulation schemes may be employed to improve the performance of the system, however this choice results in consumption of larger signal power needed to maintain the same signal separation and thus the same error probability. In a power-limited environment, the desired system performance must be achieved with the smallest possible power. One solution is the use of error-correcting codes, which increase the power efficiency by adding extra bits to the transmitted symbol sequence. However, this procedure requires the modulator to operate at a higher data rate and requires a larger bandwidth. This is essentially due to the classical approach that considers coding and modulation as two separate parts of a digital communication system. In a classical system, the information sequence is divided into message blocks of k information bits and $n - k$ redundant bits are added to each message to form a code word. The coded sequence is then modulated using one of the digital modulation techniques and fed to the channel. At the receiver part, the received signal is first demodulated, later the n bits from the demodulator corresponding to a received code word are passed to the encoder which compares the received signal with all possible transmitted code words and decides in favour of the code word, that is closest in Hamming distance (number of bit positions in which two code words differ) to the received one.

About a decode ago, using random coding bound arguments, it was shown that considerable performance improvement could be obtained by treating coding and modulation as a single entity, named as *Trellis Coded Modulation* (TCM) [1]. Its main attraction comes from the fact that it allows the achievement of significant coding gains over conventional uncoded multilevel modulation. These gains are obtained without bandwidth expansion or reduction of the effective information rate as required by traditional error-correcting schemes. TCM employs redundant non-binary modulation with a finite-state encoder which governs the selection of modulation signals to generate coded signal sequences.

A general structure of a TCM encoder is shown in Figure 4.1. In this figure, at each time i, a block of m information bits, $(a_i^1, a_i^2, \ldots, a_i^m)$ enters the TCM encoder. From these m information bits, $\tilde{m} \leq m$ bits are encoded by a rate $\tilde{m}/(\tilde{m} + 1)$ convolutional encoder into $\tilde{m} + 1$ coded bits while the remaining bits $m - \tilde{m}$ bits are left uncoded. The $\tilde{m} + 1$ output bits of the convolutional encoder are used to select one of the $2^{\tilde{m}+1}$ possible subsets of the expanded signal set and the remaining $m - \tilde{m}$ uncoded bits are used to select one of the $2^{m-\tilde{m}}$ signals in this subset. At the time i, the block $(c_i^1, c_i^2, \ldots, c_i^{m+1})$ is mapped to the signal points of the 2^{m+1}-ary signal set in such a way that the minimum Euclidean distance between channel sequences is maximized. The increase in the Euclidean distance results in a better performance when compared to that of the conventional modulation techniques.

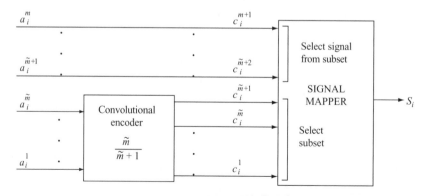

Figure 4.1 General structure of a TCM encoder

For the decoding part, the Viterbi algorithm is used to find the allowable sequence of channel symbols, that is closest in Euclidean distance to the received sequence at the channel output. The Euclidean distance between two sequences denoted by x and $\hat{x}$ of length N is given by

$$d_E(x, \hat{x}) = \sqrt{\sum_{i=1}^{N} (x_i - \hat{x}_i)^2} \tag{4.1}$$

Given x, finding the sequence $\hat{x}$ that minimizes $d_E(x, \hat{x})$ is equivalent to finding the sequence that minimizes

$$\rho_N(x, \hat{x}) = \frac{1}{N} \sum_{i=1}^{N} (x_i - \hat{x}_i)^2 \tag{4.2}$$

which is the squared error distortion measure used typically in source coding. Utilizing this analogy and noting that any set of sequences $C = \{\hat{x}_1, \hat{x}_2, \ldots, \hat{x}_k\}$, each of length N, defines a source code, the set of all allowable channel sequences and the Viterbi decoder from TCM formulation can be used as a *source code* and corresponding *source coder*. Therefore, given a data sequence x, the Viterbi algorithm is used to find the sequence $\hat{x}$ in C that minimizes $\rho_N(x, \hat{x})$.

4.3   General Aspects of Combined Trellis Coded Quantization/ Modulation Schemes

There are numerous parallels between modulation and source coding theories. Both areas mostly depend on signal space concepts and have benefited tremendously from block and trellis coding formulations. Therefore, it is possible to exploit the duality between modulation theory and source coding in order to develop novel source coding techniques. During the last two decades, trellis coded modulation (TCM) has proven to be a very effective modulation scheme for band limited channels. *Trellis coded quantization* (TCQ) [2] was introduced as a natural dual to TCM. The theoretical justification for this approach is the alphabet constrained rate distortion theory, which basically depends on the idea of finding an expression for the best achievable performance for encoding a continuous source using a finite reproduction alphabet, and this theory can be considered as a dual to the channel capacity argument, that is the motivation point for trellis coded modulation.

For the simplest form of TCQ, let us assume that for integral rate of R b/sample encoding is desired. Then 2^{R+1} quantization levels (codewords) are used, partitioned into 4 subsets, each of 2^{R-1} codewords. The subsets are used as the labels on a trellis with 2 branches entering and leaving each trellis state. For the case of $R = 3$, it takes 1 b/sample to specify the codeword within each subset, so that the encoding rate is R b/sample. The R b/sample may be thought of as a binary codeword for TCQ, and we refer to the single bit that specifies the branch as the *least significant bit* (LSB), and the remaining $(R - 1)$ bits as the *most significant bits* (MSBs). Decoding is accomplished by using LSB to specify the trellis branch, and the MSBs to specify the point by a rate—in this example—as 1/2 convolutional code, the output bits of which specify the appropriate branch subset.

Based on this analogy, trellis coded quantization (TCQ) was investigated as an efficient scheme for source coding [1–5]. The sources may be discrete or continous. For a discrete source, a specific reproduction alphabet must be chosen in order to compute the rate distortion function, while in the continous case, the reproduction alphabet is implicitly the entire real life. Alphabet constraint rate distortion theory was developed in a series of papers by W. A. Pearlman and A. Chekima [6]. The basic idea is to find an expression for the best achievable performance for encoding a continous source using a finite reproduction alphabet. The options available when choosing an output alphabet are as follows [2]:

- Choosing only the size of the alphabet (the number of elements).
- Choosing the size and the actual values of the alphabet.
- Choosing the size, values and the probabilities which the values to be used.

To explain, main source coding approaches used in TCQ, let X be a source, producing independent *independent and identically distributed* (i.i.d.) outputs according to some continous *probability density function* (p.d.f.), fx. Consider prequantizing X with a high rate scalar quantizer to obtain the source U taking values in $\{a1, a2, \ldots, aK\}$ with probabilities $P(a1), P(a2), \ldots, P(aK)$. Then encoding U as $\hat{X}$ where $\hat{X}$ takes values in $\{b1, b2, \ldots, bJ\}$. The distortion of the system is as

$$E[X - \hat{X}] = E[(U - Q - \hat{X}] = E[Q^2] + E[(U - \hat{X})^2] - 2E[Q(U - \hat{X})^2] \qquad (4.3)$$

where quantization noise is defined as $Q = U - X$. Taking the expectations,

$$E[Q(U - \hat{X})] = \sum_{k=1}^{K} \sum_{j=1}^{J} \int q(a_k - b_j) f(q|a_k, b_j) P(a_k, b_j) dq \qquad (4.4)$$

Since $f(q|a_k, b_j) = f(q|a_k)$, then Equation (4.4) can be simplified as

$$E[Q(U - \hat{X})] = \sum_{k=1}^{K} P(a_k) \sum_{j=1}^{J} (a_k - b_j) P(b_j|a_k) \int q f(q|a_k) dq \qquad (4.5)$$

For the Lloyd–Max quantizer, $E[Qa_k] = 0$, then Equation (4.6) can be rewritten as

$$= \sum_{k=1}^{K} P(a_k)(a_k - E[\hat{X}|a_k]) E[Q|a_k] \qquad (4.6)$$

$$E[(X - \hat{X})^2] = E[(X - U)^2] + E[(U - \hat{X})^2] \qquad (4.7)$$

TCQ outperforms the other source coding techniques of comparable complexity in encoding of both memoryless (e.g. uniform, Gaussian, Laplacian) and sources with memory (e.g. Gauss–Markov, sampled speech). Here, Lloyd–Max and Optimum coding methods are chosen, since their performance is higher compared to others.

TCQ and TCM can be combined in a straightforward way to produce an effective joint source coding and channel coding/modulation system. Suppose that the reproduction codebook size (i.e. number of quantization levels) for the trellis coded quantizer, is selected as $N = 2^{R+\text{CEF}}$, where $R \geq 1$ is the encoding rate in bits/sample, r and CEF are positive integers satisfying $1 \leq r \leq R$ and CEF ≥ 0. The parameter CEF stands for 'codebook expansion factor', since the codebook size is 2^{CEF} times that of a nominal R bits/sample scalar quantizer. There are totally $N_1 = 2^{r+\text{CEF}}$ subsets and N is chosen such that it can be properly divided by N_1, so each subset has exactly $N_2 = N/N_1 = 2^{R-r}$ codewords. The trellis coded quantizer maps each source sample into one of the N quantization levels by using the Viterbi algorithm. The output is a sequence of binary codewords, each of length R, with r bits to specify the subset and the remaining $R - r$ bits to determine the codeword in the specified subset. The trellis coded quantizer is followed by a TCM system which maps each output binary codeword of the source encoder into a channel transmission symbol. This mapping is one-to-one and therefore introduces no distortion. The receiver consists of a TCM decoder and a TCQ decoder. The TCM decoder maps the channel output sequence into a binary codeword sequence using the Viterbi decoding algorithm. Then, the TCQ decoder maps the binary codeword sequence into a TCQ quantization level sequence. This cascade structure of TCQ and TCM blocks gives the overall system known as *joint trellis coded quantization/modulation* (joint TCQ/TCM) system (Figure 4.2). General approach to the selection of a joint TCQ/TCM system is to assume that TCQ and TCM bit and symbol rates are equal so that the squared distance between channel sequences is commensurate with squared error in the quantization. The mapping from quantization level within a TCQ subset to modulation level within a TCM subset is selected in such a way that the level/symbol order is consistent. Since the probability of a TCM error is related to the squared Euclidean distance between the allowable paths through the trellis, a consistent labelling guarantees that Euclidean squared distance in modulation symbol space is in line with mean square error in quantization.

Joint TCQ/TCM system was introduced by M. W. Marcellin and T. R. Fischer [3] and some results were reported for the simple case when TCQ source sample rate is equal to the TCM symbol rate. In joint TCQ/TCM structure (Figure 4.2), the source is TCQ

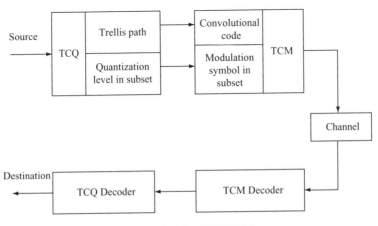

Figure 4.2   Basic joint TCQ/TCM system

encoded, creating R bit binary word for each source sample. The LSB of the TCQ output binary word is applied as the input bit to the TCM convolutional encoder. The $(R - 1)$ MSBs of the TCQ binary word then specify the modulation symbol in the TCM subset. The decoding is accomplished by first using the Viterbi algorithm in the TCM decoder, and then applying the selected R bit binary codeword as input to the TCQ decoder.

The mapping from quantization level within a TCQ subset to modulation level within a TCM subset should be selected in the obvious way, so that the level/symbol order is consistent. Since the probabiliy of a TCM error is related to the squared Euclidean distance between allowable paths through the trellis, a consistent labelling guarantees that Euclidean squared distance in the modulation symbol space is commensurate with *Minimum Squared Error* (MSE) in the quantization, hence the TCM errors of large Euclidean squared distance which cause large MSE in the source coding, will be very unlikely.

In most studies of joint source/channel coding, one of two problem formulation is used. In the first, a digital channel model is assumed (usually, *binary symmetric channel* (BSC)) and the source code is designed so that channel errors cause as little increase in distortion as possible. The second formulation to joint source/channel coding is to allow the selection of modulation symbols and the mapping from source coder levels to modulation symbols to be under the purview of the system designer.

Although their system achieves a good performance at high channel signal-to-noise ratios, the performance curves exhibit a dramatic degradation at low values due to the lack of system optimization. For a joint TCQ/TCM system, the optimization can be carried out separately for the source coding part and channel coding part or an overall system optimization can be considered. M. Wang and T. R. Fischer [4] attempted to compensate for the degradation in [3] and developed a technique for the design of TCQ/TCM systems such that the drop in the performance was largely avoided. They used a generalized Lloyd algorithm to iteratively update the TCQ levels and a quasi-Newton optimization subroutine to optimize TCM symbols, which results in, however, only locally optimal results. Later, Aksu and Salehi [5] considered channel optimized quantization levels and asymmetric signal constellations for optimum system design and proposed a simulated annealing based algorithm which finds the global optimum TCQ and TCM symbols. These methods result in 0.5–4 dB signal-to-quantization noise ratio (SQNR) gains over the non-optimized systems, which provides the gain of going to one step higher-order trellis.

The setup of joint TCQ/TCM systems in previous studies [3–5] is unnnecessarily complex. It is worth noting that the cascade structure of TCQ and TCM blocks may be renounced due to one-to-one mapping between the quantization level and the channel symbols. For instance, in the case of the codebook expansion factor is chosen as CEF $= 1$, the TCQ encoder simply generates a sequence of quantization levels from an alphabet size of $N = 2^{R+1}$ and these levels are mapped directly to symbols in the equally spaced 2^{R+1}-point TCM alphabet. Therefore, TCQ and TCM trellis structure can be combined in such a way that TCQ/TCM system operates on only one identical trellis. On the branches of the combined trellis diagram, both quantization levels and channel symbol set are placed using Ungerboeck rules. The performance of the combined trellis coded quantization/modulation, with a single trellis to describe the overall scheme, was investigated over different type of channels [7–9]. For instance, in the case of the codebook expansion factor is chosen as CEF $= 1$, the TCQ encoder simply generates a sequence of quantization levels from a codebook of size $N = 2^{R+1}$ and these levels are mapped to modulation symbols in the 2^{R+1}-point TCM signal constellation. Since there is one-to-one correspondence between the quantization level within a TCQ subset and the modulation symbol within a TCM subset, the cascade organization of TCQ and TCM blocks may be renounced. In our study, TCQ and TCM trellis structures are combined in such a way that TCQ/TCM system operates on only one identical trellis. On the branches of the combined trellis diagram, both quantization levels $q_{k,l}$ which denotes the lth level in the kth quantization subset Q_k with $k = 0, 1 \ldots N_1 - 1$, $l = 1, 2 \ldots N_2$ and signal set s_j with $j = 0, 1 \ldots N - 1$ are placed using Ungerboeck rules [1]. Thus, a single trellis is sufficient to describe the overall combined scheme under the assumption that identical trellises are used (Figure 4.3). This system is denoted as 'Combined Trellis Coded Quantization/ Modulation [7]' and has advantage over classical joint systems in terms of decoding time and complexity.

To improve Combined TCQ/TCM performance, a training sequence based on numerical optimization procedure is investigated following the Marcellin and Fischer [2] approach for output alphabet design. The principle behind training sequence design algoritm is to find a source coder that works well for a given set of data samples, that is representive of the source to be encoded. For a Combined TCQ/TCM system (trellis, output alphabet and partition) and a set of *fixed* data sequences to encode (a training set), the average distortion incurred by encoding these sequences can be thought of as a function of the output alphabet. For an alphabet of size $J = 2^{R+1}$, the average distortion is a function of J symbols in the output alphabet and so maps $\Re^J$ to $\Re$ where $\Re^J$ and $\Re$ are J-dimensional and one-dimensional Euclidean spaces. Optimization of the output alphabet can be carried out by any numerical algorithm which solves for a vector in $\Re^J$ that minimizes a scalar function of J variables. At each time the numerical algorithm updates the output alphabet estimate, the training sequences must be reencoded to compute the resulting distortion. The design process is extremely computationally intense. For this reason, the output alphabets are chosen symmetric about the origin, increasing convergence of decreasing free variables to half.

The performance of Combined TCQ/TCM with the optimized output alphabets (for encoding rates of 3 bits per sample and less) and the Lloyd–Max alphabets (for higher rates) is very good. For the simple four-state trellis, the sample average distortion is within 0.59 dB of the distortion rate function.

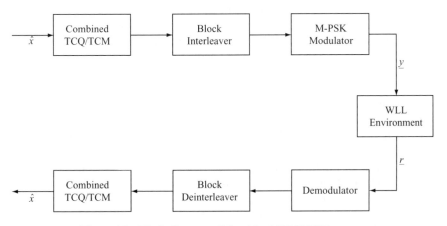

Figure 4.3   Block diagram of Combined TCQ/TCM system

4.4 Basic Model

4.4.1 Channel Model

In this tutorial, performance of Combined TCQ/TCM scheme is investigated over wireless local loop environment. In classical wired communication systems, a house is connected to a switch via first a local loop, then a distribution node. In recent years, *Wireless Local Loop* (WLL) begins to replace the local loop section with a radio path rather than a copper cable [10]. In principle, WLL is a simple concept to grasp: it is the use of radio to provide a telephone connection to the home. In practice, it is more complex to explain because wireless comes in a range of guises, including mobility, because WLL is proposed for a range of environments and because the range of possible telecommunications delivery is widening. It is concerned only with the connection from the distribution point to the house. The distribution point is connected to a radio transmitter node and a radio receiver is mounted on the side of the house. In a WLL system, wireless communication is achieved by microwave propagation [11]. The main contribution of this chapter is to demonstrate the performance of the combined trellis coded quantization/modulation system over Wireless Local Loop (WLL) environment modelled with realizable and practical medium parameters. In our wireless local loop environment model, the transmitter and receiver are point-to-point microwave links separated by a microwave channel model. Here, medium electrical parameters, i.e. dielectric constant and conductivity, vary through the channel. The considered microwave channel is shown to be Rician distributed by means of computer simulation based on Finite-Difference Time Domain technique [12,13]. The performance analysis of the combined system is carried out through the evaluation of signal-to-quantization noise ratio versus signal-to-noise ratio and bit error probability performances.

In classical wired telephone networks, a house is connected to a switch via first a local loop, then a distribution node onto a trunked cable going to the switch. Historically, the local loop was copper cable burried in the ground or carried on overhead pylons and the truncated cable was composed of multiple copper pairs. WLL replaces the local loop section with a radio path rather than a classical copper wire. Using radio rather than copper cable has a number of advantages. It is less expensive to install radio and radio units are installed only when the subscribers want the service. It is concerned only with the connection from distribution point to the house. WLL is low cost relative to deploying twisted pair or cable. It offers high-speed deployment compared to twisted pair or cable, allowing customers to be attracted before the other operators can offer them service. WLL is the use of radio to provide a telephone connection to home. WLL systems are proposed for voice, data, Internet access, TV, and other new applications of modern life.

Here, we assume that the transmitter and the receiver are point-to-point microwave links, separated by a microwave channel model. Electrical parameters such as dielectric constant ε, conductivity σ vary through the proposed channel. In our proposed WLL system, the distribution point is connected to a radio transmitter, a radio receiver is mounted on the side of the house (Figure 4.4). Mean and variance of medium parameters characterize the channel behaviour and produce the noisy environment.

For modelling the microwave channel under consideration, a computer simulation based on *Finite-Difference-Time Domain* (FD-TD) [12] technique is adopted. The Finite-difference-Time Domain (FD-TD) formulation is a convenient tool for solving

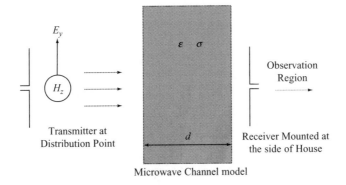

Microwave Channel model

Figure 4.4 Structure of the WLL environment

scattering problems of *Electromagnetic* (EM) fields. The FD-TD method, first introduced by Yee [12] in 1966 and later developed by Taflove [13], is a direct solution of Maxwell's time-dependent curl equations. In FD-TD, Maxwell's equations in differential form are simply replaced by their central-difference approximations, discretized and coded for numerical implementations.

In an isotropic, lossy medium, Maxwell's equations can be written as

$$\nabla x \vec{E} = -\mu \frac{\vec{H}}{\partial t} \qquad (4.8a)$$

$$\nabla x \vec{H} = \sigma \vec{E} + \frac{\partial \vec{E}}{\partial t} \qquad (4.8b)$$

The vector equation (4.8) represents a system of six scalar equations, which can be expressed in rectangular coordinate system (x,y,z) as:

$$\frac{\partial H_x}{\partial t} = \frac{1}{\mu}\left[\frac{\partial E_y}{\partial z} - \frac{\partial E_z}{\partial y}\right] \qquad (4.9a)$$

$$\frac{\partial H_y}{\partial t} = \frac{1}{\mu}\left[\frac{\partial E_z}{\partial x} - \frac{\partial E_x}{\partial z}\right] \qquad (4.9b)$$

$$\frac{\partial H_z}{\partial t} = \frac{1}{\mu}\left[\frac{\partial E_x}{\partial y} - \frac{\partial E_y}{\partial x}\right] \qquad (4.9c)$$

$$\frac{\partial E_x}{\partial t} = \frac{1}{\varepsilon}\left[\frac{\partial H_z}{\partial y} - \frac{\partial H_y}{\partial_z} - \sigma E_x\right] \qquad (4.9d)$$

$$\frac{\partial E_y}{\partial t} = \frac{1}{\varepsilon}\left[\frac{\partial H_x}{\partial z} - \frac{\partial H_z}{\partial x} - \partial E_y\right] \qquad (4.9e)$$

$$\frac{\partial E_z}{\partial t} = \frac{1}{\varepsilon}\left[\frac{\partial H_y}{\partial x} - \frac{\partial H_x}{\partial y} - \sigma E_z\right] \qquad (4.9f)$$

Following Yee's notation [12], we define a grid point in the solution region as $(i,j,k) = (i\Delta x, j\Delta y, k\Delta z)$ and any field component of space and time as

$$E^n(i,j,k) = E(i\Delta x, j\Delta y, k\Delta z, n\Delta t) \qquad (4.10)$$

where $\Delta s = \Delta x = \Delta y = \Delta z$ are the space increment, Δt is the time increment, while (i,j,k) and n are integers. Using central finite difference approximation for space and time derivatives that are second-order accurate

$$\frac{\partial E_x}{\partial y} = \frac{E_x^n(i,j+1/2,k) - E_x^n(i,j-1/2,k)}{\Delta s} \qquad (4.11a)$$

$$\frac{\partial E_x}{\partial t} = \frac{E_x^{n+1/2}(i,j,k) - E_x^{n-1/2}(i,j,k)}{\Delta t} \qquad (4.11b)$$

In applying Equation (4.11) to all the space derivatives in Equation (4.9), Yee [12] places the components of E and H about a unit cell of the lattice as shown in Figure 4.5. The computational volume of FD-TD is a space, where simulation is performed. This volume is divided into small reference cells, where the electric and magnetic fields are updated at each time step. The material of each cell within the computational volume is specified by giving its permeability, permittivity and conductivity. The material may be air (free-space) metal (perfect electric conductor) or dielectric. To incorporate Equation (4.11), the components of E and H evaluated at alternate half-time steps. Thus, to obtain the explicit finite difference approximation of Equation (4.9) first electrical field components are calculated as

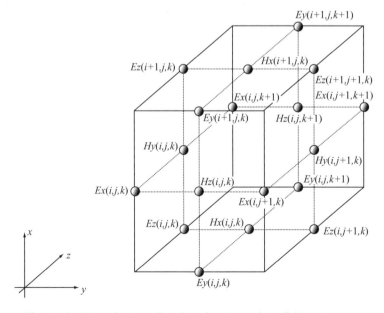

Figure 4.5   The unit Yee cell and the locations of the field components

$$E_x^{n+1}(i+1/2,j,k) = \left(1 - \frac{\sigma \Delta t}{\varepsilon \Delta s}\right) E_x^n(i+1/2,j,k) + \frac{\Delta t}{\varepsilon \Delta s}\left(H_x^{n+1/2}(i,j+1/2,k+1)\right.$$
$$- H_x^{n+1/2}(i,j+1/2,k-1/2) + H_z^{n+1/2}(i-1/2,j+1/2,k)$$
$$\left. - H_z^{n+1/2}(i+1/2,j+1/2,k)\right) \tag{4.12}$$

at a given instant in time, then the magnetic field components are found at the next instant in time.

$$H_x^{n+1/2}(i,j+1/2,k+1/2) = H_x^{n-1/2}(i,j+1/2,k+1/2) + \frac{\Delta t}{\mu \Delta s}\left\{E_y^n(i,j+1/2,k+1)\right.$$
$$\left. - E_y^n(i,j+1/2,k) + E_z^n(i,j,k+1/2) - E_z^n(i,j+1,k+1/2)\right\} \tag{4.13}$$

The calculation is repeated up to the desired observation period.

To ensure the accuracy of the computed results, the spatial increment Δs must be small compared to the wavelength (usually $\ll \lambda/10$) or minimum dimension the scatterer. To garantie the stability of the finite-difference scheme of Equations (4.12) and (4.13), the time increment Δt must satisfy the following stability condition:

$$\Delta t \leq \frac{1}{v_m \sqrt{\left[\frac{1}{\Delta x^2} + \frac{1}{\Delta y^2} + \frac{1}{\Delta z^2}\right]}} \tag{4.14}$$

Where v_m is the maximum wave phase velocity within the computational domain. For practical reasons, it is best to choose the ratio of the time increment to spatial increment as large as possible yet satisfying Equation (4.14).

A basic difficulty encountered in applying the FD-TD method to the scattering problem is whether the domain in which the field to be computed is open or unbounded. Since no computer can store an unlimited amount of data, a finite-difference scheme over the whole domain is impractical. We must limit the extent of our solution region. In other words, an artifical boundary must be enforced to create the numerical illusion of an infinite space. The solution region must be large enough to enclose the scatterer, and suitable boundary conditions on the artificial boundary must be used to simulate the extension of the solution region to infinity. Outer boundary conditions of this type have been called either radiation conditions, absorbing boundary conditions, or lattice truncation conditions.

For the planar wave propagation, we consider one-dimensional wave propagation. Assume waves have only E_z and H_x components and propagete in the $\pm y$ directions. Also assume a time step of $\Delta t = \Delta y/c$, the maximum allowed by the stability condition of Equation (4.14). If the lattice extends from $y = 0$ to $Y = J\Delta y$, with E_z component at the end points

$$E_z^n(J) = E_z^{n-1}(J-1) \tag{4.15}$$

and the truncation conditions are

$$E_z^n(0) = E_z^{n-1}(1) \tag{4.16}$$

With these lattice conditions, all possible $\pm y$ directed waves are absorbed at $y = 0$ and $J\Delta y$ without reflection. The initial field components are obtained by simulating either an incident plane wave pulse or modulated plane wave. The simulation should not take excessive storage nor cause spurious wave reflections. A desirable plane wave source condition takes into account the scattered fields at the source plane. A typical wave source condition at plane $y = js$ is as

$$E_z^n(js) = E_z^{n-1}(js) + y(t) \tag{4.17}$$

where $y(t)$ is the modulated signal.

In simulation studies, 10^6 Combined TCQ/TCM 8-PSK modulated signals are passed through the microwave channel and FD-TD numerical computational methods are used to compute the field values within and out of the channel medium. In this method, time-domain differential Maxwell equation is discretized by using the numerical differences. By the utilization of FD-TD, real time field distribution can be calculated over environment channel with realistic physical medium parameters. In our model, we assume that the incident electromagnetic wave has only y electrical component, $E_y(x, t)$ and z magnetical component, $H_z(x, t)$ and wave propagation is in x direction as shown in Figure 4.4.

As a result of the computer simulation based on FD-TD, fading parameter ρ of the WLL environment is found to be a function of dielectric constant ε, conductivity σ and the ratio of the channel model width to wave length (d/λ) as follows:

$$\rho = f(\varepsilon, \sigma, d/\lambda) \tag{4.18}$$

In our case, wavelength is equal to 3 cm due to the fact that 10 GHz is used as carrier frequency. The simulations are done for the slab width 4 and 7 cm. The simulation results obtained for different SNR values (Figure 4.6) also show that the p.d.f. of ρ matches to Rician p.d.f. with fading parameter $K = 5$ and $K = 10$ dB for the slab width $d = 4$ and 7 cm respectively . It is also shown that for the same dielectric constant ε, conductivity σ and wavelength λ, as channel width d decreases, the Rician channel parameter increases. Rician p.d.f. can be written as

$$p(\rho) = 2\rho(1 + K) \exp\left(-K - \rho^2(1 + K)\right) I_0(2\rho\sqrt{K(1 + K)}) \quad \rho \geq 0 \tag{4.19}$$

where ρ is fading coefficient, K is fading parameter and I_0

$$I_0(x) = \sum_{k=0}^{\infty} \left[\frac{x^k}{2^k k!}\right]^2 \tag{4.20}$$

is the zero order modified Bessel function of first kind. The parameter K represents the ratio of average power in the direct (line off sight) and specular component to that in the scattered component. The Rayleigh fading corresponds to the limiting case of Rician fading channel when K is equal to 0 (no direct or specular component) and its probability density function is expressed as

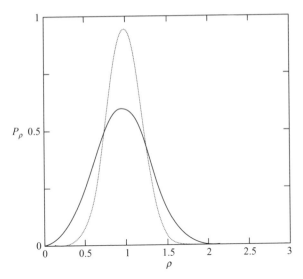

Figure 4.6 Normalized probability density function of the WLL environment Dashed lines ($K = 5\,\mathrm{dB}$), Solid Lines ($K = 10\,\mathrm{dB}$)

$$p(\rho) = 2\rho \exp(-\rho^2) \quad \rho \geq 0 \tag{4.21}$$

Since in WLL systems, the fading effects can be modelled as being Rician distributed [9], we can state that our proposed model matches perfectly to WLL environment.

4.4.2 System Model

The main emphasis in this chapter is to analyse the combined TCQ/TCM system over WLL system modelled with microwave channel, which is investigated in the previous section, for the special case of codebook expansion factor is chosen equal to unity. The basic system under consideration (Figure 4.3) accepts as input real continuous amplitude, discrete-time source sequence produced by a memoryless Gaussian source.

Combined TCQ/TCM encoder converts the source sequence of length L, $\underline{x} = (x_1, x_2, \ldots, x_L)$ into a sequence of encoder output symbols, which are then block interleaved to break up burst errors caused by amplitude fades of duration greater than one symbol time. In practice, the interleaving depth should be chosen on the order of maximum fade depth anticipated. Here, the infinite interleaving depth is assumed for the purpose of analytical approach since this assumption provides a memoryless channel for which well-known bit error probability upper bounding techniques can be used.

At i^{th} signalling interval, the interleaved symbol is mapped into the M-PSK signal where M is given as $M = 2^{R+1}$. Corresponding to the M-PSK symbol sequence $\underline{y} = (y_1, y_2, \ldots, y_L)$, a noisy discrete-time sequence $\underline{r} = (r_1, r_2, \ldots, r_L)$ appears at the output of the channel. The received signal at i^{th} signalling interval is expressed as

$$r_i = \rho_i \cdot y_i + n_i \tag{4.22}$$

where n_i is the additive Gaussian noise and ρ_i is Rician distributed due to the structure of microwave channel under investigation. At the receiver, first the noise-corrupted sequence is demodulated and deinterleaved. Later it is passed through the Combined TCQ/TCM decoder which employs Viterbi algorithm to determine the most likely coded symbol sequence transmitted and produces the output sequence of quantization levels $\hat{x} = (\hat{x}_1, \hat{x}_2, \ldots, \hat{x}_L)$ under the assumption that there is one-to-one mapping from quantization level within a TCQ subset to modulation symbol within a TCM subset.

Throughout the chapter, signal-to-quantization noise ratio (SQNR) is adopted as the performance measure

$$\text{SQNR} = \sum_{i=1}^{L} E\left[x_i^2\right] \Big/ \sum_{i=1}^{L} E\left[(x_i - \hat{x}_i)^2\right] \tag{4.23}$$

The distortion rate function evaluated at the channel capacity provides an upper bound to the overall SQNR performance, that is possible. If $D(R)$ is the distortion rate function for the source, then the *optimum performance theoretically attainable* (OPTA) is derived by substituting the channel capacity C at the given signal-to-noise ratio in the source distortion rate function $D(R)$. In the case of Gaussian source is used, which is also under investigation here, the SQNR performance is upper bounded by

$$\text{OPTA} = 10 \cdot \log_{10}\left(\frac{1}{2^{-2C}}\right) \tag{4.24}$$

For the derivation of OPTA curves, it is necessary to define the channel capacity for the considered channel. In AWGN environment, the channel capacity is well known to be

$$C = \frac{1}{2}\log_2(1 + \gamma) \text{ bits/symbol} \tag{4.25}$$

where γ is the signal-to-noise ratio as $\gamma = S/BN_0$ with the carrier power S, channel bandwidth B and the noise spectral density $N_0/2$. For the computation of channel capacity over Rayleigh fading channel, which models the microwave channel under investigation, we adopt the model introduced by W. C. Y. Lee [14] and developed in [15].

The channel capacity for fading channels must be calculated in average sense. Since fading results in a variation of the above bandlimited AWGN channel on a slower time scale, the instantaneous capacity given by Equation (4.25) has to be averaged over the relevant ensemble. The parameter subject to change is the carrier power or equivalently the signal-to-noise ratio. The average signal-to-noise ratio Γ is expressed as

$$\Gamma = \rangle\gamma\langle = \frac{\rangle S\langle}{BN_0} \tag{4.26}$$

Denoting p_Γ as the probability density function associated with Γ, the average channel capacity for the fading environment is given W. C. Y. Lee and C. G. Gunther by [14,15]

$$C = \frac{1}{2}\int_0^\infty \log_2(1 + \gamma)p_\Gamma(\gamma)\,\mathrm{d}\gamma \tag{4.27}$$

Altough OPTA curve for AWGN case may be viewed as an overall upper bound for SQNR performance, the derivation of channel capacity for Rician fading channel lets us to define the expilict upper bound for the considered fading environment.

Besides the SQNR performance, the bit error probability of the considered system is also investigated. It has been shown that, for AWGN channels and for $\frac{n}{n+1}$ encoder, an upper bound on the average bit error probability P_b, assuming ideal interleaving/deinterleaving is obtained by union bound as [16]

$$P_b \leq \frac{1}{n} \frac{dT(D,I)}{dI}\bigg|_{I=1, D=\exp(-E_b/4N_0)} \tag{4.28}$$

where E_b is the average power of the modulated signal sequence and N_0 is one sided noise power. $T(D,I)$ is the transfer function of the pair-state transition diagram that takes into account an enumaration of all distance and error properties associated with the trellis code, and with the number of states equivalent to the number of encoder states, squared. Unfortunately this approach does not seem to be practical for analyzing trellis codes with more than two states. In a broad class of trellis codes, referred to as *uniform*, identical for which the encoder transfer function can be obtained from a modified state state transition having no more states than the encoder itself. The trellis codes employed in our model are chosen from the class of uniform trellis codes.

In the presence of fading, evaluation of P_b depends on the proposed decoding metric, the presence or absence of the *channel state information* (CSI) and the types of detection used. In the case of coherent detection, ideal CSI with an additive maximum likelihood metric and ideal interleaving/deinterleaving, it has been shown that an upper bound on the average bit error probability P_b, is as [16]

$$P_b \leq \frac{1}{n} \frac{d\overline{T(D,I)}}{dI}\bigg|_{I=1} \tag{4.29}$$

where n is the number of source bits per trellis transition and $\overline{T(D,I)}$ is the modified transfer function of the error state diagram whose branch labels differ from AWGN case as follows. In the absence of fading each branch label gain has a factor D^{β} where β represents the squared Euclidean distance between any two channel symbols. In our case, we simply replace D^{β} by $\overline{D^{\beta\rho^2}}$ which is given by

$$\overline{D^{\beta\rho^2}} = \frac{1+K}{1+K+\beta\zeta} e^{\frac{\beta K}{1+K+\beta\zeta}}, \quad \zeta = \frac{E_b}{4N_0} \tag{4.30}$$

where K is fading parameter and E_b/N_0 is the average bit energy to noise.

4.5 An Example : 4-State 8-Psk Combined Trellis Coded Quantization/Modulation

In this section, as an example a four state combined TCQ/TCM structure is investigated over WLL environment. We select $r = CEF = 1$, so the reproduction codebook is partitioned into $N_1 = 4$ subsets, each with exactly $N_2 = 2^{R-1}$ codewords. The encoding rate is chosen as $R = 2$ bits/sample, however the method can be straightforwardly extended to higher rates.

The system consists of a four state Combined TCQ/TCM scheme employing an 8-PSK constellation. On the branches of the proposed combined structure, there is one-to-one correspondence between the signal set and quantization levels (Figure 4.7). Two adjacent branches, each of which contains two parallel transitions, emanate from each state. On the transitions of the trellis diagram, quantization levels, $q_{k,1} \in Q_k$, $k = 0, 1, 2, 3$; $l = 1, 2$ and modulation symbols s_j, $j = 0, 1 \ldots 7$ are placed as in Figure 4.8. From every state emanates two adjacent branches each of which contains two parallel transition. The resulting combined trellis code can easily be shown to satisfy the conditions of uniformity. The modified state diagram (Figure 4.9) is presented in the form of a flow graph, commencing and terminating in all zero state. Each branch label has been augmented with a term I^l (I is an abstract parameter) where l corresponds to the number of information bits which differ in the binary information vectors, associated with the all zero's state transition, and the particular state transition under consideration.

SQNR performance for the considered system is investigated through the computer simulation. For each SNR value, 10^5 samples which are generated from a memoryless

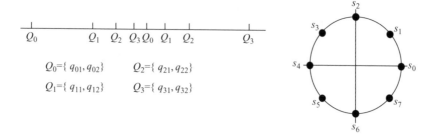

$Q_0 = \{q_{01}, q_{02}\}$ $Q_2 = \{q_{21}, q_{22}\}$

$Q_1 = \{q_{11}, q_{12}\}$ $Q_3 = \{q_{31}, q_{32}\}$

Figure 4.7 Quantization levels and Signal constellation

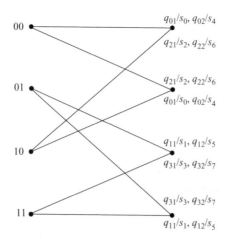

Figure 4.8 The trellis diagram for 4-state 8-PSK Combined TCQ/TCM scheme

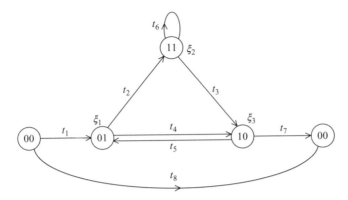

Figure 4.9   Error state diagram for 4-state 8-PSK Combined TCQ/TCM scheme

Gaussian source of zero mean and unit variance, are processed by the four state 8-PSK Combined TCQ/TCM system. In Figure 4.10, SQNR performance curves are illustrated for microwave channel in both cases of Lloyd–Max and Optimum [2] quantization levels being employed. The utilization of Optimum quantization levels improves the performance compared to that obtained when Lloyd–Max levels are used. It is also shown that simulation results are upper bounded by the OPTA curve derived for Rician fading.

Besides the SQNR performance, the considered system is also investigated through the bit error probability. In order to evaluate the bit error performance, transfer function approach is adopted. The error state diagram for the scheme is shown in Figure 4.10 with branch gains

$$
\begin{aligned}
t_1 &= b \cdot I + b \cdot I^2 \\
t_2 &= 0.5 \cdot (a + c) \cdot I + 0.5 \cdot (a + c) \cdot I^2 \\
t_3 &= 0.5 \cdot (a + c) \cdot I + 0.5 \cdot (a + c) \cdot I^2 \\
t_4 &= a + c \cdot I \\
t_5 &= 1 + d \cdot I \\
t_6 &= a + c \cdot I \\
t_7 &= b \cdot I + b \cdot I^2 \\
t_8 &= d \cdot I
\end{aligned}
\tag{4.31}
$$

where

$$
a = \overline{D^{0.58\rho^2}}, \; b = \overline{D^{2\rho^2}}, \; c = \overline{D^{3.41\rho^2}}, \; d = \overline{D^{4\rho^2}}
\tag{4.32}
$$

are given by Equation (4.29).

The modified transfer function is found by solving the below equations, derived from the error state diagram in Figure 4.10, simultaneously.

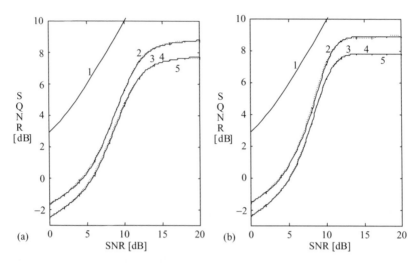

Figure 4.10 SQNR performance over WLL environment: and (a) for $K = 5$ dB, and (b) for $K = 10$ dB. (1) Opta Curve (2) Optimum Analytical Curves (3) Optimum Simulation Curves (4) Lloyd–Max Analytical Curves (5) Lloyd–Max Simulation Curve

$$\xi_1 = t_1 + \xi_3 t_5$$
$$\xi_2 = \xi_1 t_2 + \xi_2 t_6$$
$$\xi_3 = \xi_2 t_3 + \xi_1 t_4 \tag{4.33}$$
$$\overline{T(D,I)} = \xi_3 t_7 + t_8$$

Consequently, $\overline{T(D,I)}$ is obtained as

$$\overline{T(D,I)} = \frac{1}{1 - t_4 t_5} \left[\frac{t_1 t_2 t_3 t_7}{(1 - t_4 t_5)(1 - t_6) - t_2 t_3 t_5} + t_1 t_4 t_7 \right] + t_8 \tag{4.34}$$

Replacing the modified transfer function in Equation (4.29), the analytical upper bounds are derived for WLL environment under investigation in both cases of Lloyd–Max and Optimum quantization levels are used. The effect of quantization noise for both levels are taken into account in Figure 4.11 under the consideration that the quantization process results in extra power consumption.

In order to verify the analytical results, we also performed computer simulations in the presence of Rician fading, which models the microwave channel. The random variables representing the normalized fading amplitude at each signalling interval were generated from the Rician probability distribution, statistically independent of each other so as to satisfy our assumption that interleaving depth be infinite. Also, the decision depth for the Viterbi algorithm was chosen to be equal to 10 signalling interval, which yields an acceptable delay for all practical purposes. As shown in Figure 4.11, the simulation curves confirm the analytical results.

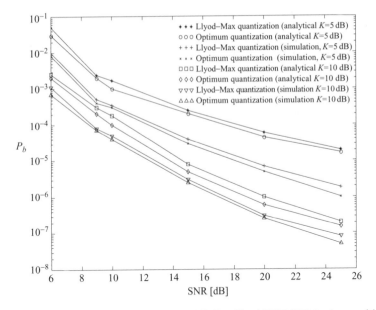

Figure 4.11   Analytical and simulation curves of 8-PSK Combined TCQ/TCM scheme with Lloyd–Max and optimum quantization levels over WLL environment

References

[1] G. Ungerboeck, 'Channel Coding with Multilevel/Phase Signals,' *IEEE Trans. Inform. Theory*, vol. 28, no. 1, pp. 55–67, 1982.

[2] M. W. Marcellin and T. R. Fischer, 'Trellis Coded Quantization of Memoryless and Gauss–Markov Sources,' *IEEE Trans. Commun.*, vol. 38, no. 1, pp. 82–93, 1990.

[3] M. W. Marcellin and T. R. Fischer, 'Joint Trellis Coded Quantization/Modulation,' *IEEE Trans. Commun.*, vol. 39, no. 2, pp. 172–176, 1991.

[4] M. Wang and T. R. Fischer, 'Trellis Coded Quantization Designed for Noisy Channels,' *IEEE Trans. Inform. Theory*, vol. 40, no. 11, pp. 1792–1802, 1994.

[5] H. A. Aksu and M. Salehi, 'Joint Optimization of TCQ-TCM Systems,' *IEEE Trans. Commun.*, vol. 44, no. 5, pp. 529–533, 1996.

[6] W. A. Pearlman and A. Chekima, 'Source Coding Bounds using Quantizer Reproduction Levels,' *IEEE Trans. Inform. Theory*, vol. 30, pp. 559–567, 1984.

[7] M. Uysal and O. N. Uçan, 'Combined Trellis Coded Quantization/Modulation over Fading Mobile Channel,' *Proceedings of ACTS Mobile Communication Summit '97*, Aalborg, Denmark, pp. 906–911, 7–10 October 1997.

[8] O. N. Uçan, M. Uysal and S. Paker, 'Performance of Combined Trellis Coded Quantization/Modulation over Microwave Channel,' in *Internationa Conference on Telecommunications (ICT)*, Thessaloniki, Greece, June 1998.

[9] O. N. Uçan, M. Uysal and S. Paker, 'Performance of Combined Trellis Coded Quantization/Modulation over Wireless Local Loop Environment,' *International J. Commun. Systems, Special Issue on Wireless Local Loop*, accepted for publication.

[10] D. J. van Wyk, M. P. Lotter, L. P. Linde and P. G. W. van Rooyen, 'A Multiple Trellis Coded Q^2PSK Systems for Wireless Local Loop Applications,' in *Proc. IEEE International Symposium on Personal, Indoor and Mobile Radio Communications (PIMRC)*, pp. 624–628, 1997.

[11] W. Webb, *Introduction to Wireless Local Loop*, Artech House, Boston-London, 1998.

[12] K. S. Yee, 'Numerical Solution of Initial Boundary Value Problems Involving Maxwell's Equation in Isotropic Media,' *IEEE Trans. Anten. Propagat.*, vol. 14, no. 5, pp. 302–307, 1966.

[13] A. Taflove, *Computational Electrodynamics: the Finite Difference-Time Domain Method*, Artech House, London, 1995.

[14] W. C. Y. Lee, 'Estimate of Channel Capacity in Rayleigh Fading Environment,' *IEEE Trans. Vehicular Tech.*, vol. 39, no. 8, pp. 187–189, 1990.

[15] C. G. Gunther, 'Comment on Estimate of Channel Capacity in Rayleigh Fading Environment,' *IEEE Trans. Vehicular Tech.*, vol. 45, no. 5, pp. 401–403, 1996.

[16] D. Divsalar and M. K. Simon, 'Trellis Coded Modulation for 4800–9600 bits/s Transmission over a Fading Mobile Satellite Channel,' *IEEE J. Selected Areas Commun.*, vol. 5, no. 2, pp. 162–175, 1987.

5

Low Sequency W-CDMA Codes Lead to More Economic WLL and Infostation Terminals

James G. Evans and B. R. Badrinath

5.1 Introduction

Wireless Local Loop (WLL) systems offer quick deployment, high data rates and dynamic capacity. These should be compelling reasons for WLL System to supplant wired systems. WLL systems are having only limited success in displacing wired systems because wired systems provide proven reliability at lower cost. Will this always be true? This chapter addresses that one technology path designers can follow to reduce WLL system cost, especially for the more numerous low data rate and voice user. A discussion of cost and other market needs is helpful to understand the technical challenges that must be addressed to have WLL systems compete more successfully with wired alternatives.

5.1.1 WLL Market Challenges

In developed nations there is a large demand for additional connectivity because of the need for Internet access, a growth of small businesses in the home and the demands of children for their own phone. Examining the United States as a market, there are approximately seven million businesses and 110 million homes [1]. Of the businesses, over 80 % have less than 20 employees. Thus, there is a large small office (the applicability of WLL to large businesses is unclear. There certainly is an opportunity for in-building wireless systems), home office (SOHO) and residential market with increasing communication needs. This is a cost sensitive market with voice being the primary communication need and data needs rapidly growing. These needs are primarily served today by the *Local Exchange Carriers* (LECs). The current method to add capacity is a second line or *local loop* (LL). The average embedded cost of the LEC local loop is $600. An emerging alternative is to install *Asymmetric digital Subscriber line* (ADSL) technology that multiplexes additional communications capacity above the voice signal on one line. The cost is approximately $300 for the customer premises ADSL modem, $100 for installation, and additional cost for a PC interface card and matching equipment at the telephone company office. Again the cost is not too different than the $600 traditional loop, albeit this technology is capable of much higher data rates. Cable TV operators and cable TV operators in combination with long distance telephone companies (e.g. AT&T + TCI) are addressing 'the last mile' to the SOHO market by converting the one-way cable TV systems to two-way. The conversion cost is $400–600 per customer for and incumbent operator and several times more for a start-up system [2].

This brief look at competing technologies establishes approximately $600 as a target WLL cost for the SOHO market in the US and other developed countries. The primary customer need at this target cost is voice and low rate data (e.g. 64 Kbps). Of course many customers will not be satisfied with low data rates and a WLL system will not be sold if it is not capable of simultaneously serving all customers but not all at the above target cost.

For simplicity assume that this $600 WLL target cost for the low-end customer is equally divided between the radio *Network Interface Unit* (NIU) at the end customer's premises and the radio base station of the service provider. Thus, an installed NIU should cost $300. A WLL system is complicated and a manufacturer or supplier will incur a considerable design and sales overhead. For this situation the 'rule of thumb' for manu-facturing cost is one third of the installed cost. This sets a manufacturing cost target of less than $100 for the radio NIU. The radio in this NIU, the primary focus of this chapter, should be less than half of this target. Is this an impossible goal?

One can look at the consumer electronics market to see that this target cost for a radio is achievable. In the US the retail price for 25 W marine radio is $100, albeit it uses analogue modulation. The retail price for a digital cordless phone, two radios, is also approaching $100. Yet, using another 'rule of thumb' where the manufactured cost of a consumer product is half of the retail price, one can see that there is hope in reaching the above target costs for the NIU, if the designers make the right technology choices. Picking wisely is what this chapter is about.

In developing countries the market needs are similar to the above with a greater emphasis to provide low-end voice service. Where no communications are available, all technologies will compete on an equal footing for start up systems. Power may be unreliable and battery back up operation for extended periods is then important. This makes low power consumption a design requirement. Although high data rate enhanced services and mobile services may not be needed at first, a WLL system that cannot grow into these services will not be bought. Often systems are selected based upon local manufacturing content. This places the more complex WLL technology at a disadvantage and therefore cost is all that more important.

While wired LL systems may be more competitive in delivering high data rates, WLL systems have an advantage in economically providing mobile services. A service like *Personal Handy Phone* (PHS) in Japan could easily be offered through the WLL base stations. Another mobile service possibility is the Infostation concept being researched at Rutgers University. An Infostation [3] is a small wireless cell where a mobile user can get Internet access at low to high data rates [4] at a very low cost per bit. The Infostation concept includes simple terminals for short messages through to high data rate terminals for large file exchanges. The simple terminals must be inexpensive, small and have a small battery and yet operate in a common environment with more capable terminals. This is the same requirement for WLL systems.

In summary the technical challenge is to design a single radio system that simultan-eously supports high power, high data rate terminals, down to mass market NIUs and miniature mobile terminals that operate for a long time on a small battery. All this must be done at a cost competitive with wired alternatives.

5.1.2 Research Objective and W-CDMA

This chapter addresses the challenge to implement a common physical layer, specifically the radio, and the base-band circuits that will allow megabit data rates to complex terminals and not require unnecessary cost, processing power, and battery size in simpler terminals. Multirate *Wide-band Code Division Modulation* (W-CDMA) is selected for the physical radio layer because it has many of the attributes needed for wireless systems to be more competitive in the above-mentioned markets. A common radio [often costly and inflexible circuits] in the base station can simultaneously serve low and high data rate customers. Furthermore, W-CDMA modulation and demodulation processes are compatible with a 'software' implementation. This offers flexibility in data rates, BER, or range. *The most significant finding is a new way in selecting and processing long and short codes that allows the customers requiring only low data rates (e.g. voice) to use less signal processing and thereby achieve longer battery life and lower cost.*

This work focuses on the low data rate receivers in the base to terminal radio link. In a WLL or an Infostation system, these receivers must be low in cost and continuously powered to avoid 'ring' delay for a voice circuit. Long periods of battery back up operation are a design requirement. Mobile Infostation terminals must continuously receive because a cell can be traversed in a few seconds in a vehicle. The continuous powering limits the technology choices to avoid costly batteries.

This base to terminal link (forward link) is where orthogonality between codes can easily be maintained for multiuser and multirate access. Furthermore, this link is challenging because it must support higher data rates than the reverse link; e.g. the home Internet and the mobile user is typically the recipient and not the source of information. These concepts apply to the reverse link provided the transmissions are synchronized, power is controlled, and other complexities are addressed. A discussion of these complexities is beyond the scope of this chapter. This research is an extension of the third-generation mobile radio research of K. Okawa and F. Adachi [5].

5.1.3 Other Technologies

The remaining sections of this chapter will focus on W-CDMA as a good choice to achieve the WLL objectives of flexibility in data services and low cost for the low-end customer. Other radio technologies were considered.

Time division multiple access (TDMA) technology [Examples: DECT, GSM] was not chosen because the low-end user that may be satisfied with the capacity of an occasional time slot still has to process signals at the full system data rate. The radio circuits have to be full bandwidth, equalize the channel and be capable of high peak power. These requirements are not compatible with low cost.

Orthogonal frequency division multiplexing (OFDM), where data are modulated on hundreds of narrow bandwidth RF carriers is intriguing. A base station could modulate a high data rate signal on a many carriers and many time slots. The low data rate user could be modulated on a few carriers and time slots. This low data rate signal may be inexpensive to generate and receive, especially at the base station. OFDM has the disadvantage that the many carrier RF signal has a high peak to average power ratio. This requires very linear and consequently power inefficient and expensive RF circuits. Therefore, this technology is not attractive in the end user WLL or Infostation terminals.

W-CDMA technology remained as the most attractive choice to achieve the objectives of flexibility and low cost. Data rates can be changed by straightforward Boolean operations in the base band circuits. Binary modulation of the RF signals can be used and thereby achieve RF signals with low peak to average power ratios. The reader is encouraged to explore other RF technologies, perhaps hybrids of the above, to have radio solutions be successful in the market place.

5.2   Code Selection and Generation

A significant challenge in designing a multiuser and multirate W-CDMA system is generating codes that are orthogonal. This orthogonality prevents interference from one user into another. An algorithm to generate attractive codes is presented below.

Let $\underline{C}_m = \{C_i(t)\}$, $i = 1, 2, \ldots m$ denotes a matrix of spreading codes that are orthogonal over the time interval $0 \leq t \leq \tau$. τ is the duration of the symbol for the highest data rate user. Let n denote the chip length of each of these codes. n is selected to achieve the necessary spreading or processing gain for the highest data rate user to satisfy the communications link budget. m is selected to have enough codes to serve all users as explained below. It may be desirable to assign several codes to one user (multicode) to achieve an even higher data rate. For simplicity, a multicode user is treated as additional single code users.

$$\underline{C}_m = \begin{vmatrix} C_1(t) \\ C_2(t) \\ C_3(t) \\ \vdots \\ C_m(t) \end{vmatrix} \tag{5.1}$$

The code words in C_m are ordered in terms of increasing number of state transitions per time interval, referred to as sequency [6]. This is a metric related to occupied bandwidth, a property used later. Without loss of generality, understanding can be enhanced by considering a specific example where C_m is constructed by reordering codes from a binary Hadamard matrix H_n, $m \leq n$. A Hadamard matrix H_n is an $n \times n$ matrix of $+1$s and -1s such that the inner product of any pair of distinct code words is 0. [This can be implemented through a multiplication and summation or the binary exclusive or operation and summation.] Hadamard matrices have many useful properties [7]. E.g. $H_n H_n T = nI$, $H_n T = nH_n - 1$ and the n rows of H_n supplemented by their complements form $2n$ code words of length n with a minimum Hamming distance of $n/2$. Furthermore, if H_n is a Hadamard matrix then by the Sylvester Construction technique, shown below, so is H_{2n} where

$$\underline{H}_{2n} = \begin{vmatrix} H_n & H_n \\ H_n & -H_n \end{vmatrix} \tag{5.2}$$

Consider the 16×16 Sylvester constructed and reordered Hadamard matrix shown below. The ordering is the number of state transitions. A transition at the beginning and end of the code word is assumed with a probability equal to 1/2. Therefore, the first word has one transition and the second two and the last 16.

$$
\underline{C}_{16} =
\begin{array}{l}
1111111111111111 \\
11111111\text{-}\text{-}\text{-}\text{-}\text{-}\text{-}\text{-}\text{-} \\
1111\text{-}\text{-}\text{-}\text{-}\text{-}\text{-}\text{-}\text{-}1111 \\
1111\text{-}\text{-}\text{-}\text{-}1111\text{-}\text{-}\text{-}\text{-} \\
11\text{-}\text{-}\text{-}\text{-}1111\text{-}\text{-}\text{-}\text{-}11 \\
11\text{-}\text{-}\text{-}\text{-}11\text{-}\text{-}1111\text{-}\text{-} \\
11\text{-}\text{-}11\text{-}\text{-}\text{-}\text{-}11\text{-}\text{-}11 \\
11\text{-}\text{-}11\text{-}\text{-}11\text{-}\text{-}11\text{-}\text{-} \\
1\text{-}\text{-}11\text{-}\text{-}11\text{-}\text{-}11\text{-}\text{-}1 \\
1\text{-}\text{-}11\text{-}\text{-}1\text{-}11\text{-}\text{-}11\text{-} \\
1\text{-}\text{-}1\text{-}11\text{-}\text{-}11\text{-}1\text{-}\text{-}1 \\
1\text{-}\text{-}1\text{-}11\text{-}1\text{-}\text{-}1\text{-}11\text{-} \\
1\text{-}1\text{-}\text{-}1\text{-}11\text{-}1\text{-}\text{-}1\text{-}1 \\
1\text{-}1\text{-}\text{-}1\text{-}1\text{-}1\text{-}11\text{-}1\text{-} \\
1\text{-}1\text{-}1\text{-}1\text{-}\text{-}1\text{-}1\text{-}1\text{-}1 \\
1\text{-}1\text{-}1\text{-}1\text{-}1\text{-}1\text{-}1\text{-}1\text{-}
\end{array}
\tag{5.3}
$$

(-1) is abbreviated as $(-)$ above

A binary symbol stream, $S(t) = 10110\ldots$ is spread by code word $C_i(t)$ by transmitting C_i, $-C_i$, C_i, C_i, $-C_i$, $\ldots$ A simplified schematic of this transmitter is shown below.

$S(\omega) * C_i(\omega)$, the convolution of $S(\omega)$ and $C_i(\omega)$, is the frequency domain representation of the signal out of the multiplier. $C_i(t)$ is a periodic code word generator with period τ synchronized with the binary symbols in the data signal $S(t)$. Assuming a large amount of spreading (i.e. large processing gain), the bandwidth of $C_i(\omega)$ is much larger than the bandwidth of $S(\omega)$. Therefore, the bandwidth of $S(\omega) * C_i(\omega)$ is slightly larger than the bandwidth of $C_i(\omega)$. Compare the output signals $S(\omega) * C_{16}(\omega)$ and $S(\omega) * C_2(\omega)$. Since $C_{16}(\omega) = C_2(\omega/8)/8$ the bandwidth of $S(\omega) * C_{16}(\omega)$ is approximately 8 times the bandwidth of $S(\omega) * C_2(\omega)$. These differences in bandwidth are exploited later in this chapter.

The next step is to generate code words with different spreading and inversely proportional data rates. Consider the illustrative example of generating p_1 code words of length n chips to spread a data signal with data rate r and symbol period τ, p_2 code words of length $n2^{k_2}$ to spread a rate $r/2^{k_2}$ and period $2^{k_2}\tau$ signal, and p_3 code words of length $n2^{k_2+k_3}$ to

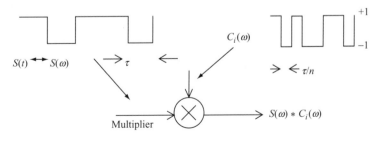

Figure 5.1 Transmitter

spread a rate $r/2^{k_2+k_3}$ and period $2^{k_2+k_3}\tau$ signal. The codes in p_1, p_2 and p_3 can be computed dynamically or fixed a priori in the system design. The following algorithm can generate these code words.

Pick p_1 code words among the highest sequency code words from $\underline{C}_m$. These code words are to be used for the p_1 users at the highest data rate r. Form a matrix $\underline{C}_{n\times(m-p_1)}$ with $m - p_1$ rows of n chips selected from the remaining rows of $\underline{C}_m$. Use the Sylvester construction to form the matrix $\underline{H}_{2n\times2(m-p_1)}$. Note that this is not a square matrix. Reorder this matrix with rows increasing in the number of state transitions to form $\underline{C}_{2n\times2(m-p_1)}$. Repeat these steps k_2 times. Each step doubles the number of orthogonal code words, doubles the chips per word and doubles the length of the code word.

$$\underline{C}_m \longrightarrow \underline{C}_{n \times (m-p_1)} \longrightarrow \underline{H}_{2n \times 2(m-p_1)} \longrightarrow \underbrace{\underline{C}_{2n \times 2(m-p_1)}}_{k_2 \;\; \text{times} \nearrow} \longrightarrow \underline{C}_{\alpha \times \beta} \qquad (5.4)$$

$C_{\alpha\times\beta}$ is a $\beta = 2k2(m - p1)$ row (orthogonal code word) by $\alpha = 2k2n$ column (number of chips) matrix. From the $2k2(m - p1)$ code words select $p2$ among the highest sequency codes for modulating $p2$, rate $r/2k2$ and period $2k2 \;\square\square\square$ data signals.

The final set of p_3 code words can now be generated. Starting with a matrix formed by removing the above selected p_2 code words from $\underline{C}_{\alpha\times\beta}$ perform k_3 operations like those symbolically represented by Equation (5.4) to generate $\underline{C}_{\alpha 2^{k_3} \times 2^{k_3} (\beta-p_2)}$. This matrix contains $2^{k_3}(\beta - p_2)$ rows or code words of $n2^{k_2+k_3}$ columns or chips. From the remaining code words select p_3 code words of length $n2^{k_2+k_3}$ to spread rate $r/2^{k_2+k_3}$ and period $2^{k_2+k_3}\tau$ data signals.

The following inequality must be met:

$$p_1 + p_2 2^{-k_2} + p_3 2^{-k_2-k_3} \le m \qquad (5.5)$$

If Equation (5.5) is satisfied without equality then the groupings of code words (p_1, p_2 and p_3) and the selection of code words within a group are not unique and additional criterion can be used for their selection.

A very important property of the above algorithm for constructing multirate codes is that all codes are orthogonal. This is proved as follows. All of the codes in the group p_1 are orthogonal to the codes in p_2 and p_3. This is because the codes in p_2 and p_3 are constructed from concatenating codes that are orthogonal to the codes in p_1 over every symbol period τ and therefore over $2^{k_2}\tau$ and $2^{k_2+k_3}\tau$. For the same reason all codes in p_2 and p_3 are orthogonal. The codes may not be orthogonal if the implicit timing coherence is not maintained. (See below for additional comments on timing coherence.)

Additional attributes of the above algorithm for selecting multirate codes will be discussed in the next section.

5.3 Infostation Transmitter

A very simplified block diagram of a WLL or an Infostation transmitter is shown in Figure 5.2.

The quantity $S_i(\omega) * C_i(\omega)$ is the frequency domain representation of the data signal $S_i(t)$ spread by code $C_i(t)$; see Figure 5.1. All signals are linearly combined and up converted

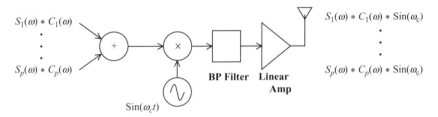

Figure 5.2 Infostation transmitter

by a radio carrier ω_c for transmission. The band pass filter must have a bandwidth larger than the bandwidth of the widest spread data signal. (This filter must be all pass and linear phase for the same reasons discussed in Section 5.5.)

5.4 Terminal Receiver

The receiver is more complex than the transmitter in a W-CDMA radio. Therefore, simplifying the receiver can be important in reducing radio cost. Since the receiver must be on all or most of the time in a WLL or Infostation system reducing power consumption has a large impact on battery life. For these reasons the remaining discussions are on receiver design.

A simplified receiver with all spread data signals at the antenna is shown in Figure 5.3. The through the air path is assumed to be free of multipath effects so as to avoid the discussion of equalization or combining RAKE receivers. Also, for simplicity, one stage of heterodyne conversion and one pre-decorrelation filter is shown [8]. A combination of pre-decorrelation (e.g. fixed filters at RF and IF) and post decorrelation filters is used in practical systems. The pre-decorrelation filters reduce interference and false synchronization. The post decorrelation filtering can be implemented with digital processing and can be 'software defined'. The objective of this receiver is to detect the data signal $S_i(t)$ and reject all other signals destined for other receivers. An estimator for the data sent, $S_i(t)$, is generated by a correlator where the received signal $S_i(\omega) * C_i(\omega)$ is despread by multiplying it with a receiver generated and time synchronized code $C_i(\omega)$. All other signals are rejected because their codes are orthogonal to $C_i(\omega)$ as explained above. The demodulator, through operations such as low-pass filtering and integrating over the symbol period, decides what binary signal was sent during each symbol period. So far this is a conventional W-CDMA receiver.

A modern receiver design would realize the above decorrelation and demodulation operations using an A/D converter and a digital signal processor of some form (e.g. combination of DSP, FPGA, and ASIC). This architecture could have the highly desirable feature of being software configurable. For example data rates and codes could be dynamically changed. It is illuminating to estimate the processing needed to implement a software configurable digital receiver.

If the highest user data rate were 1 Mbps and each 1 μsec symbol were spread 16 times then the received W-CDMA signal would have a 16 Mbps chip rate. Assume each chip is sampled 1–4 times [9] by a 12-bit A/D. Also assume 30 'instructions' (the average number of serial and parallel operations) of signal processing per sample for chip

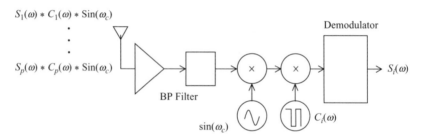

Figure 5.3   A receiver with all spread data signals at the antenna

timing, frame synchronization, equalization, CRC calculation, etc. This 'digital signal processor' would have to have a capability greater than 1000 million instructions per second (MIPs). At 1–2 mw/MIP this signal processor would dissipate 1–2 watts. This power dissipation is excessive for a small portable Infostation terminal or for a WLL voice terminal operating on battery backup (hand held GPS receivers process a 1.023 Mchip/s spread spectrum signal. Even though the designs use highly specialized ASICs, these receivers operate for only a day on several AA batteries.). The designer of these terminals would have to make design trade-offs. The most significant of these trade-offs would be to restrict operation to low data rates (another trade-off is to use more ASICs in the design and limit software reconfiguration). Assuming that a low end and low cost Infostation terminal or a terminal for a voice circuit is constrained to operate at the low data rates, *can a receiver as discussed above be implemented with less power and at lower cost?*

5.5   Simplified Signal Processing for a Low Data Rate Terminal

The receiver can be simplified if the low sequency codes generated by the above algorithm are assigned to the low data rate users. These codes have less bandwidth and they have a lower effective chip rate. These properties can lead to substantial reductions in circuit complexity and speed provided the terminal operates only at the lower data rates.

The first simplification is to reduce the bandwidth of the pre-decorrelation filter shown in Figure 5.3 to be commensurate with the bandwidth of the low data rate signal spread with a low sequency code. This filter will reject most of the energy in the signals for other users that are at the higher data rates. This filter also will reject other interfering signals that are outside of the filter bandwidth. Consequently, the range of signal strengths over which the receiver RF and IF must remain linear is greatly reduced. This is especially important when there are several base stations and the low data rate user might be closer to an adjacent base station than its own, the 'classic CDMA near far problem'.

This filter, or combination of filters, must produce an all pass, linear phase, transfer function to preserve the orthogonality between codes. This requirement is shown as follows: The code $C_i(t)$ is a periodic signal with a period $T = 2^{k_2+k_3}\tau$ or $2^{k_2}\tau$ or τ in the example. Therefore, a Fourier series can represent $C_i(t)$.

$$C_i(t) = \sum_{q=-\infty}^{\infty} c_{iq} e^{j\omega qt}, \quad \omega = 2\pi/T \tag{5.6}$$

The cross correlation between $C_i(t)$ and $C_k(t)$ is

$$\rho_{ik} = 1/T \int_0^T C_i(t) C_k(t) dt$$

$$= \sum_{q=-\infty}^{\infty} c_{iq} * c_{kq} = 0 \text{ for } \neq k \tag{5.7}$$

$$= 1 \text{ for } i = k$$

If $C_i(t)$ is limited in bandwidth, e.g. a bandwidth passing the main lobe of the $(\sin \omega/\omega)^2$ spectrum,

$$C_i(t) \sim \sum_{q=-\Gamma}^{\Gamma} c_{iq} e^{j\omega qt} \tag{5.8}$$

i.e. $c_{iq} \sim 0$ for $|q| > \Gamma$ and

$$\rho_{ik} = \sum_{q=-\infty}^{\infty} c_{iq} * c_{kq} \sim \sum_{q=-\Gamma}^{\Gamma} c_{iq} * c_{kq} = 0 \quad \text{for } i \neq k \tag{5.9}$$

The amplitude of each term in Equation (5.9) is unaltered by an all pass transfer function. The phase of each product is unchanged because the linear phase produces a delay that the timing circuits in the receiver cancel. Therefore, orthogonality is preserved even though $C_k(t)$, possibly a wider bandwidth code, undergoes the narrow bandwidth filtering that just passes $C_i(t)$.

Table 5.1 below presents two illustrative system designs derived from the above concepts. The 1024 Mbps, highest data rate signals, are spread 16 times. A receiver would require an equivalent low-pass bandwidth of slightly more than 16.384 MHz and 16.384 Mcps processing. The lowest data rate receivers could be implemented with equivalent low-pass bandwidths of slightly more than 4.096 MHz and 2.048 MHz in the two systems. The respective chip processing speeds would be 4.096 Mcps and 2.048 Mcps, a 4:1 and 8:1 reduction in processing speed (and battery drain for this function) compared to the high data rate receiver.

From these examples it is apparent that the number of codes and therefore the number of users at the lower data rates is limited. Yet, these are practical capacities for small cell Infostation and WLL systems. An unconstrained system with a 16.384 Mbps channel rate would support 256 simultaneous users at 64 Kbps. Of course all receivers would have to process the 16.384 Mbps signal at great cost.

Table 5.1   The two system designs

# Users	Data Rate Kbps	Effective Spreading Ratio	Effective Chip Rate Mcps	Total Data Capacity Mbps
System 1				
	1024	16:1	16.384	8.192
8				
	256	32:1	8.192	4.096
16				
	64	64:1	4.096	4.096
64				
				16.384
System 2				
	1024	16:1	16.384	8.192
8				
	256	32:1	8.192	6.144
24				
	32	64:1	2.048	2.048
64				
				16.384

5.6   Noise Analysis and Processing Gain

An important subject is the analysis of noise. The common assumption in a W-CDMA radio is that noise and interference into the decorrelator is reduced by the ratio of the chip rate to the data rate. This is referred to as processing gain. Since the system noise and broadband interference bandwidth is increased by this same ratio, one breaks even with perfect decorrelation. Therefore, W-CDMA systems offer no information theoretic capacity advantage over other modulation techniques. This common wisdom on processing gain can be verified and some interesting properties of Hadamard codes can be developed at the same time through the analysis that follows.

Let $N(t)$ represent the band-limited noise at the input to the decorrelator of Figure 5.3. Over the data symbol time interval $[0, T]$, this noise can be approximated by

$$\bar{N}(t) = \sum_{i=0}^{n-1} N(iT/n)\, U_n(t, iT/n) \tag{5.10}$$

The unit pulse function $U_n(t, s)$ is defined in Figure 5.4. $U_n(t, s) = 1$ in the shaded areas and equals 0 otherwise.

$\bar{N}(t)$ is obviously a good estimator for $N(t)$ for sufficiently large n if $N(t)$ is bounded and Reimann integrable. Following the logic of J. L. Massey [10], n should be just large enough so that $\bar{N}(t)$ is a good estimator of $N(t)$ over n dimensional Euclidean space. The mean square error for finite n, assuming $N(t)$ is a stationary process with autocorrelation function $R(t)$, is

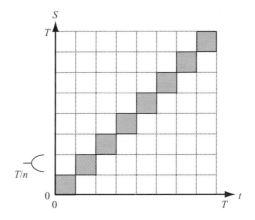

Figure 5.4   The unit pulse function $U_n(t,s)$

$$E[N(t) - \bar{N}(t)]^2 = 2\left[R(0) - \sum_{k=0}^{n-1} R(t - kT/n) U_n(t, kT/n) \right]$$
$$= 2[R(0) - R(\varepsilon)] \tag{5.11}$$

$\varepsilon = t$ modulo T/n. $\bar{N}(t)$ is not a stationary estimator. The mean square error is small if the digital decorrelator samples its input signals at $t = iT/n$, $i = 1, n$. The precise sampling time is arbitrary since $N(iT/n)$ in Equation (5.10) could be replaced with $N(iT/n + \delta)$ and the remaining equations would be unchanged. A mid chip sampling (i.e. $\delta = T/2n$) is a logical choice.

A striking property of the Hadamard waveforms in the matrices $\underline{H}_n$ or $\underline{C}_n$ is

$$U_n(t, s) = 1/n \sum_{i=1}^{n} C_i(t) C_i(s) \tag{5.12}$$

$C_i(t)$, $i = 1, n$ are Hadamard codes, n chips long, over the period $t = 0, T$. A direct consequence of Equation (5.11) is that $C_i(t)$ are the eigenvectors (a.k.a. characteristic functions) of the kernel $U_n(t, s)$ and the eigenvalues are 1. A substitution of Equation (5.12) into Equation (5.10) yields the following:

$$\bar{N}(t) = \sum_{i=0}^{n-1} N(iT/n)\left[1/n \sum_{j=1}^{n} C_j(t) C_j(iT/n) \right]$$
$$= \sum_{j=1}^{n} \left[1/n \sum_{i=0}^{n-1} N(iT/n) C_j(iT/n) \right] C_j(t)$$
$$= \sum_{j=1}^{n} a_j C_j(t) \tag{5.13}$$

The third line in Equation (5.13) should not be a surprise because the waveforms $C_j(t)$, $j = 1, n$, are complete in n dimension Euclidean space.

The coefficients $(a_j, j = 1, n)$ are random variables of zero mean and cross correlation:

$$E[a_j a_p] = 1/n^2 \sum_{i=0}^{n-1} \sum_{k=0}^{n-1} [R(iT/n - kT/n) C_j(iT/n) C_p(kT/n)] \tag{5.14}$$

If $N(t)$ is a process formed by ideal low-pass filtering of white noise[i], then $R(\tau) = N_o \sin(\omega_o \tau)/2\pi\tau$. N_o is the 'one sided' spectral density of the white noise process and $\omega_o = 2\pi f_o$ is the cut-off frequency of the low-pass filter. If $T/n = 1/f_o$ then $R(iT/n) = 0$ for $i \neq 0$. [Another property [8] is that the random variables $N(iT/n)$ in Equation (5.10) are independent.] Then $E[a_j a_p] = R(0)/n = N_o f_o/n$ for $j = p$ and 0 otherwise. Thus, the random variables $(a_j, j = 1, n)$ are independent and identically distributed (i.i.d.) (if $E[N(t)N(s)] = R(t, s) = U_n(t, s)$ then, by the Karhunen–Loeve Theorem, $C_j(t), j = 1, n$ are the characteristic functions of $R(t, s)$ and Equation (5.12) is and orthogonal representation. The coefficients a_j are i.i.d. and Equation (5.11) equals 0. The above allows a calculation of signal-to-noise ratio.

At the input to the decorrelator shown in Figure 5.3 the signal can be represented as

$$\sum S_j C_j(t) + \sum_{j=1}^{n} a_j C_j(t) \tag{5.15}$$

The first sum represents all signals and the second the noise (and interference) passed by the receiver RF and IF filters. The decorrelator multiplies the composite signal in Equation (5.15) with the code $C_i(t)$ assigned to user 'i'. Due to the orthogonal properties of the codes this process results in only signal S_i and noise a_i being detected with a signal to noise

$$S_i^2/E[a_i^2] = S_i^2/N_o f_o/n \tag{5.16}$$

This signal to noise is the ith signal power and $1/n$ of the total noise power passed by these filters. Thus, this orthogonal code representation of noise reveals the $n{:}1$ processing 'gain' inherent with W-CDMA.

References

[1] US Census.
[2] A. Dutta-Roy, 'Bringing Home the Internet,' *IEEE Spectrum*, vol. 36, no. 3, pp. 32–38, March 1999.
[3] R. H. Frenkiel and T. Imielinski, 'Infostations: The Joy of "Many-time, many-where" Communications,' WINLAB Technical Report TR-119, Apr. 1996.
[4] D. Goodman, J. Boras, N. B. Mandayam and R. Yates, 'Infostations: A New System Model for Data and Messaging Services,' *VTC 97*, Phoenix, AZ, May 1997.
[5] K. Okawa and F. Adachi, 'Orthogonal Multi-Spreading Factor Forward Link for Coherent DS-CDMA Mobile Radio,' *Proc. IEEE ICUPC 97*, pp. 618–622, San Diego, USA, Oct. 12–14 1997.

[6] H. Zhang and D. Rutkowski, 'Orthogonal Sequency Division Modulation—A Novel Method for Future Broadband Radio Services,' *Proc. of 45th IEEE Vehicular Technology Conference (VTC-95)*, pp. 810–814, July 1995.

[7] S. B. Wicker, *Error Control Systems for Digital Communication and Storage,* Prentice-Hall, Englewood Cliffs, NJ, Chapter 6, 1995.

[8] R. C. Dixon, *Spread Spectrum Systems With Commercial Applications*, Third Edition, John Wiley & Sons, New York, pp. 144, 172, 264, 1994.

[9] T. Kirke, 'Interpolation, Resampling, and Structures for Digital Receivers,' *Communication Systems Design*, pp. 43–49, July 1998.

[10] J. L. Massey, 'Information Theory Aspects of Spread-Spectrum Communications,' *International Symposium on Spread Spectrum Theory and Applications*, pp. 16–21, 1994.

[11] A. Papoulis, *Probability, Random Variables, and Stochastic Processes*, McGraw-Hill Book Company, New York, p. 372, 1965.

6

Wide-band Wireless Outdoor to Indoor Local Loop Channel Models for Urban and Suburban Environments at 2 GHz

Ian Oppermann and Jaakko Talvitie

6.1 Introduction

Wireless communications systems offer a great deal of flexibility for mobile users and represent a key component in future global personal communications. Local loop systems are envisaged to allow the integration of many services into a high-speed digital system which allows communication inside buildings and access to the external world. Wireless systems potentially offer unprecedented flexibility for home or office data communications systems and reduced overheads costs for both installation and equipment. For reasons associated with both cost and flexibility, the idea of WLL, or 'wireless to the home' has generated significant interest with telecommunications service providers. Some systems have been proposed that it should be possible to offer low mobility users very high data rates and may even be intergrated into B-ISDN services [1].

In order to assist in the development of such systems, knowledge of the characteristics of the WLL channel must be investigated. To date, little has been published on wide-band WLL channel models at this frequency, so little is known about the channel characteristics. Narrowband channels are far better understood [2,3] and narrowband WLL systems have proven the usefulness of the approach. Recent papers published which address the subject of wide-band WLL channel characteristics and has measurement details are given in [4,5].

Therefore, to determine the characteristics of a wide-band channel model for the WLL in urban and suburban environments, a series of measurements were carried out in offices and homes both at Sydney, Australia and in Helsinki, Finland. Seperate measurement procedures were followed in Sydney and Helsinki, however similar information was obtained at the two locations.

The experiments consisted of taking *impulse response* (IR) measurements in many locations corresponding to typical office environments and suburban homes. The responses include measurements taken using both directional and omni-directional transmitting antennas in a number of locations. The effect of people moving near the receiver was also considered.

For the purposes of this paper, a WLL system is defined to be one in which the receiver is stationery or moving at pedestrian mobility. The short term characteristics of the channel were investigated by taking successive measurements for several seconds in many locations in the WLL environment. Issues such as path loss, however, were not addressed.

The parameters extracted from the data which are used to characterize the channel include the power-weighted, *root-mean-square* (RMS) delay spread of the IR [6], the *carrier to multipath ratio* (CMR), and the number of MPCs. Values for each of these parameters have been calculated for each location where measurements were taken and are presented in tables. Average statistics for these parameters were also calculated based upon rooms measured, distance examined and each propagation scenario. The distribution of the amplitudes of the first signal component and the most significant MPCs are also investigated.

The extracted parameters of the measured IR were used to simulate the channel as a tap-delayed transversal filter with time-varying parameters. In order to model the environment, it has been assumed that the channel is *wide-sense stationary consisting of uniform scatters* (WSS-US). Furthermore, ergodicity of the channel is assumed. The static characteristics of the channel observed in the measurements support these assumptions. The close fit of channel characteristics achieved using the model based on these assumptions also supports the WSS-US and ergodicity views.

The remainder of this paper is set as follows. Section 2 describes the experimental procedure used in the two measurement campaigns, Section 3 describes the data processing stages used to determine the channel parameters, while Section 4 describes the extracted channel parameters. Section 5 describes the channel model used and finally conclusions are presented in Section 6.

6.2   Experimental Procedure

As mentioned in the introduction, two different measurement techniques were used for the measurement campaigns. This was essentially due to the availability of equipment at each research centre. In one location, time-domain measurements were taken directly, while at the other, measurements were made in the frequency domain and then converted to the time domain.

6.2.1   Frequency Domain—(Sydney)

The measurement procedure, described in [7], consisted of a stationary receiver taking frequency domain, transfer function readings at various distances and angles from a stationary transmitter (base station). Unlike the Hashemi method, however, the separations between antennas (200–500 m) meant that standard RF cables could not be employed. As a result, a high-speed fibre optic link was used to remote the transmit antenna. The advantage of a fibre optic link for this application is the low loss of the cable which is also bandwidth independent. The link consisted of a Fabry–Perot laser diode at 1300 nm which was intensity modulated at approximately 1.8 GHz by an electro-optic modulator. The RF source for the modulation on the electro-optic modulator is derived from the *S*-parameter test set of the vector *network analyser* (NWA). The modulated light was delivered to the transmit antenna via a length of ruggedized single-mode fibre. Single-mode fibre has extremely low loss (0.02 dB/km) and low dispersion for the frequency range of interest. A high-speed photo-detector converted the light-wave signal back to an electrical signal, which was then amplified by two cascaded microwave amplifiers, resulting in a signal level of approximately + 22 dBm at the antenna input.

A remote link to the receiver (user terminal) was also used for some parts of the experiment. This consisted of a directly modulated laser diode which was connected to the microwave receiver by a single-mode optical fibre. At the receiver, a high-speed photo-detector was again used to convert the modulated light-wave signal to an electrical signal, which was then amplified and delivered to the network analyser for the measurement of the transfer function of the channel.

The receiving antenna considered consisted of an omni-directional monopole antenna. Two transmit antennas were considered; a highly directional horn antenna and an omni-directional monopole antenna (see Figures 6.1 and 6.2). For each combination of antennas and base station position, on average five rooms were measured under both conditions of controlled movement and under conditions of no movement of people near the receive antenna.

In this paper, a *measurement scenario* refers to a particular transmitter or simulated base station location in a particular propagation environment (urban or suburban). A *measurement location* refers to a target room where measurements are taken. A *receiver position* refers to a single point where the receive antenna will be positioned. A *measurement run* refers to a series of 128, 256 or 512 individual, consecutive recordings carried out at a given *receiver position*.

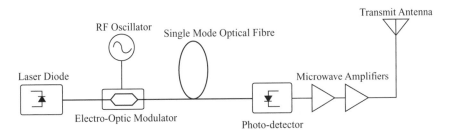

Figure 6.1 Block diagram transmit antenna connection

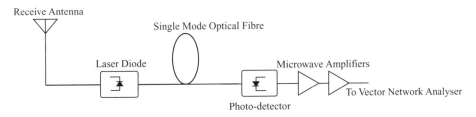

Figure 6.2 Block diagram receive antenna connection

This measurement signal itself consisted of a frequency sweep between 1.77 and 1.85 GHz in 401 steps for STILL measurements, and 101 steps for MOVE-type measurements. The time required to obtain each STILL profile was 80 ms making it possible to take up to 12 measurements per second. Each MOVE-type measurments took approximately 50 ms making it possible to take up to 20 measurements per second. The reduction in resolution for the MOVE measurements allowed an increased channel sampling rate, but as may be seen, the channel sampling rate did not increase linearly with the reduction in resolution.

Measurements were carried out for cases where there was no movement of the receiver and no movement in the vicinity of the receiver and, in cases where there was controlled movement of the receiver and movement in the vicinity of the receiver. The broad categories of measurements are listed in Table 6.1. These different types of measurements were carried out in the same position so that movement was the only variable. All movement is with respect to the receiving antenna. The limited number of cases where the receiver was moved does not lead to statstically reliable results, so the detailed results have not been listed in this paper.

The overall measurement system parameters are given in Table 6.2 below.

For the urban environment, four measurement scenarios were considered. One of the base station locations considered both the monopole transmitter and the directional antenna. The positioning attempted to represent

- A 'city street' with pole mounted omni-directional antenna.
- A 'city street' with pole mounted directional antenna.
- *One roof-top location.* The omni-directional antenna was mounted on the roof of a building of height approximately 7–8 storeys.
- *One roof-top location.* The directional antenna was mounted on the roof of a building with path distance greater than the roof-top scenario described in the point above. The roof top height was approximately 5–6 storeys.

The receiver was located inside rooms in the neighbourhood of the transmitter. In no case was there ever a *line-of-sight* (LOS) condition. Walls, doors or buildings always obstructed the direct path between transmitter and receiver. The receiver was always at a height of 1.2 m but measurements were taken on several floors of the buildings considered in at least some of the measurement scenarios.

For the 'suburban' environment, two measurement scenarios were considered. The transmit antenna was mounted on the wall of a building at a height of approximately three storeys and only the omni-directional antenna was used. The measurement locations were in the rooms of private homes and rooms in an area which is a typical inner city neighbourhood consisting almost entirely of double-storey terrace houses. The larger

Table 6.1 Types of measurements considered

Type	Rx move.	Move. immed. near Rx	Move. near Rx
STILL	No	No	No
MOVE 1	No	No	Yes
MOVE 2	No	Yes	Yes

Table 6.2 Sydney measurement system parameters

Parameter	Value
Centre frequency	1.81 GHz
Measurement bandwidth	80.0 MHz
Frequency resolution	200 kHz (401 points)
Max excess delay (STILL)	5.0 μs
Max excess delay (MOVE)	1.26 μs
Maximum Doppler freq. (STILL)	± 6.0 Hz
Maximum Doppler freq. (MOVE)	±10.0 Hz
Transmit antenna	Horn with gain > 10 dB
Transmit antenna	Monopole antenna
Receive antenna	Monopole antenna

building on which the transmitting antenna was located was on the boundary of this neighbourhood.

6.2.2 Time Domain—(Helsinki)

The measurement system is based on analogue sliding correlation shown in Figure 6.3. The transmitter modulates a carrier with an *m*-sequence. The received signal is cross-correlated with a reference signal modulated by the same sequence at a slightly lower chip rate, causing the two sequences to slide past each other. The process of sliding correlation produces consecutive estimates of the channel IR, scaled in time by a time-scaling factor depending on the sliding rate. The complex IR estimates are sampled and stored for later processing.

The transmitter and the receiver are clocked by phase-coherent rubidium frequency standards, which allows both the amplitude and the phase of the IR to be extracted. However, no provision for measuring the absolute transmission delays is made in the system. Only the excess delay values with respect to a chosen reference can be extracted from the data.

The transmitter (base station) was placed at varying heights of rooftops depending on the scenario being investigated. The receiver (user terminal) locations were situated in rooms on several floors of the buildings considered. For each transmitter location, a number of receiver locations were examined. The transmitting and receiving antennas considered were omni directional.

The measurement signal itself consisted of a 1023-chip *m*-sequence modulated to a 2.1 GHz carrier and transmitted at 1 W. The main parameters of the measurement are presented in Table 6.3. The parameters result in approximately 12 IR to be recorded per second.

Measurements were entirely of the MOVE1 type and made in rooms and buildings in suburban environments as well as down-town Helsinki. The measurements in the

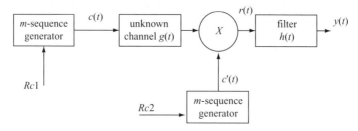

Figure 6.3 Baseband impulse response measurement system using sliding correlation

Table 6.3 Helsinki measurement system parameters

Parameter	Value
Chip frequency	53.85 MHz
Sequence length	1023 chips
k-factor	1077
Maximum Doppler frequency	± 6.1 Hz
Sampling freq.	200 kHz
Delay range	4.0 µs
Delay resoln.	19 ns
Meas. per reading	256

suburban environment were made with the transmitter mounted on poles with height equal to approximately three storeys. The receiver locations were typically first and second floor rooms in small apartment block typical of the area. Both LOS and *non-line-of-sight* (NLOS) conditions existed between the transmitter and the receiver. Two base station positions using omni-directional transmit and receive antennas were considered.

The 'metropolitan' measurements were made with the transmitter located on the 13th floor of a hotel in the centre of Helsinki. Two base station positions (hence two measurement scenarios) were considered on this floor. Receiver locations were typically selected to be the fourth or fifth floor of apartment blocks.

6.3 Data Processing

For the data from Sydney, it was first necessary to convert the results into the time domain. Using similar data processing methods as used in [7], the frequency-domain information obtained from the NWA was windowed using a three term Blackman–Harris window [8] before taking the discrete inverse Fourier transform to obtain time-domain IR. The frequency step size of 200 kHz results in a maximum measurable time delay window of approximately 5 μs. The resolution of the IR, equivalent to the inverse of the bandwidth swept, is 12.5 ns. The actual resolution of the final time-domain response is reduced by the additional width of the inverse transform of the window function, which for the 401 element window is 30 ns or approximately three samples in the time domain [7,8].

The resultant complex-valued low-pass equivalent time-domain IR may be represented in the classical form of

$$h(t) = \sum_{k=0}^{N} a_k B(t) * \delta(t - t_k) \exp^{j\theta_k} + n(t) \tag{6.1}$$

where N is the number of multipath components, $\{a_k\}, \{t_k\}$ and $\{\theta_k\}$ are the random amplitude, propagation delay and phase sequences, respectively. The $n(t)$ term in Equation (6.1) represents the low-pass, complex-valued additive Gaussian noise. The parameter $B(t)$ represents the shape of the band-limited received pulse, at positions specified by $\delta(t - t_k)$, which deviates significantly from an ideal delta function. For the Sydney measurements, this parameter is a sinc function convolved with the inverse transform of the Blackman–Harris window, while for the Helsinki measurements, the pulse shape is determined by the receiver input low-pass filter and the correlation peak of the transmitted m-sequence.

Once the Sydney IRs were converted to the time domain, all data could be treated in the same manner. Firstly, the noise floor was removed and the *multipath components* (MPCs) identified. The noise floor was removed through the use of a sliding 'matched' filter window. The window shape used was the same as that of the band-limited, delta function $B(t)$. This approach applies unequal weighting to each point in the window. If the power of the received signal inside the window exceeded some constant times the RMS value of the noise power estimate, then the central signal point inside the window was considered to be signal. Otherwise, it was deemed to be Gaussian noise. The estimate for the noise power was taken from the latest part of the measurement where little signal was expected. This approach is similar to that used in [9,10] and is equivalent to satisfying the condition

$$P_{\text{window RMS}} > P_{\text{RMS thresh}} \tag{6.2}$$

where

$$P_{\text{RMS thresh}} = k P_{\text{noise RMS}} \tag{6.3}$$

The value of the constant k was chosen empirically to minimize 'false MPC detections' but to allow sufficient sensitivity to detect most of the MPCs. The noise estimate was calculated for each individual IR.

Figure 6.4 shows an IR with the noise floor removed using the windowing approach, respectively. Using the conventional threshold approach, the achievable measurement range is restricted by the peaks of large noise samples. Figure 6.4 shows that a greater dynamic range is achievable using the windowing technique.

Since the measured channel IRs were relatively static, a further technique was employed to minimize the number of false samples detected as signal using the noise-windowing technique. If an MPC is truly present, then it should be present in several consecutive IRs. If a given sample detected as signal did not correspond to samples in at least some of the previous or later measured IRs at the same location, it was considered noise and removed.

Once an estimate of the AWGN had been removed, the individual 'resolvable' MPCs were identified. This was achieved by selective identification and removal of the largest MPCs. Identification of large MPCs relies on the knowledge of the band-limited ideal impulse $B(t)$ defined in Equation (6.1).

Since each MPC has a characteristic shape, a template with this shape is fitted to the IR magnitude and each large MPCs identified is subtracted from the IR magnitude. With some tolerance for numerical and measurement accuracy, anything remaining after the large MPC is removed must be due to other MPCs. The same treatment is then applied to the magnitude of the residue of the IR.

This technique requires the centre of the MPC to be estimated from the samples that indicate the position of a given MPC. To improve the accuracy of this approach, fractionally spaced 'impulse templates', with resolutions higher than the received signal sampling rate, were used. These were correlated with the received magnitude until a maximum value was reached. The scaled template was then subtracted from the received signal. An example of an IR which has had noise removed and MPCs detected is shown in Figure 6.5. In this figure, the circles show the position and amplitude of the detected MPCs.

This technique has the distinct advantage of allowing very closely spaced MPCs to be identified. If two MPCs are separated by one sample, and less than approximately 8 dB, then they will be resolved. Once an identified impulse is removed, a margin of error is allowed to account for finite resolution of the measurements and the corresponding

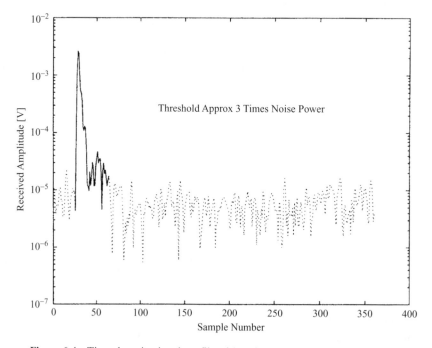

Figure 6.4 Time-domain signal profile with noise removed by sliding window

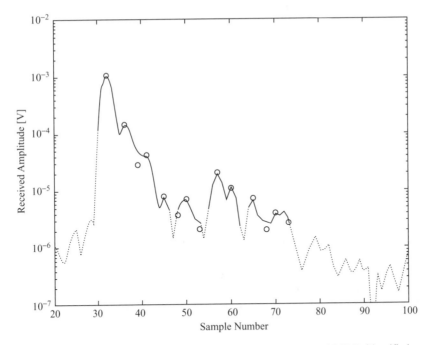

Figure 6.5   Time-domain signal profile with noise removed and MPCs identified

uncertainty in MPC location. If the residual signal at any excess delay exceeds this margin for error, the process examines the residue signal for further impulses.

The resolution available in the measurements and the data processing techniques used were sufficient to enable the resolution of the most significant MPCs. Confidence in this statement is based on the very small amplitude variations (over time) of the identified MPCs (implying little in the way of unresolved MPCs) and a typically close fit obtained when an attempt was made to reconstruct the power decay profile for a given IR based on the location of identified MPCs. The reconstruction was achieved by placing scaled copies of the band-limited ideal MPC at locations corresponding to identified MPCs.

6.4   Channel Parameters

Once the MPCs have been identified, it is then a straightforward matter to extract the parameters that describe the channel. Results for channel parameters are presented which are distinguished according to propagation scenario as described in Table 6.4. The Helsinki measurements are divided into HELSUBURBMO and HELURBMO for sub-urban and urban, MOVE1, omni-directional transmit antenna, respectively. In a similar way, the Sydney suburban measurements are divided into SYDSUBMO and SYD-SUBSO, corresponding to MOVE and STILL measurements using the omni-directional transmitter. The urban measurements are divided into SYDURBMD, SYDURBMO, SYDURBSD and SYDURBSO for MOVE directional and omni-directional antennas, and STILL directional and omni-directional antennas.

Table 6.4 Measurement scenarios

Scenario name	Location	Suburban	*Tx* antenna	Move. immed. near *Rx*	Move. near *Rx*
HELSUBURBMO	Helsinki	Yes	Omni	No	Yes
HELURBMO	Helsinki	No	Omni	No	Yes
SYDSUBMO	Sydney	Yes	Omni	Yes	Yes
SYDSUBSO	Sydney	Yes	Omni	No	No
SYDURBMD	Sydney	No	Directional	Yes	Yes
SYDURBMO	Sydney	No	Omni	Yes	Yes
SYDURBSD	Sydney	No	Directional	No	No
SYDURBSO	Sydney	No	Omni	No	No

6.4.1 Carrier to Multipath Ratio and RMS Delay Spread

One of the parameters of interest is the carrier to multipath ratio denoted by CMR. The CMR represents the relative power of the largest component to the total power of the MPCs excluding the largest component. For each received IR, the parameter is defined as

$$\mathrm{CMR} = \frac{a_l^2}{\sum_{k=0,\,k\neq l}^{N} a_k^2} \tag{6.4}$$

where a_l is the amplitude of the largest signal component and a_k is the amplitude of the kth received multipath component. Owing to the nature of the propagation environment, which is not necessarily the case, the first component has the largest received magnitude component.

One of the most important parameters to characterize the channel is the power weighted RMS delay spread of the received IR. This parameter indicates the susceptibility of the channel to *inter-symbol interference* (ISI) and ideally should be as small as possible [6,11]. The RMS delay spread for a single impulse profile is defined as

$$\tau_{\mathrm{rms}} = \sqrt{\frac{\sum_k a_k^2 (t_k - \tau_m - t_A)^2}{\sum_k a_k^2}} \tag{6.5}$$

where t_A is the arrival time of the first path in the profile and τ_m is the mean excess delay defined as

$$\tau_m = \frac{\sum_k a_k^2 (t_k - t_A)}{\sum_k a_k^2} \tag{6.6}$$

The mean, standard deviation, the maximum and minimum values for each of these parameters as well as the MPC counts are given in Tables 6.5 and 6.6 for the Helsinki and Sydney measurements, respectively. The number of measurements used to extract the

Table 6.5 Channel parameters from Helsinki measurements

Name	HELSUBURBMO	HELURBMO
Num Prof	24064	8448
Mean MPC	21.3	26.8
Std MPC	3.1	3.4
Max MPC	60	200
Min MPC	2	1
Mean CMR	3.5	0.5
Std CMR	1.9	0.4
Max CMR	164.2	15.5
Min CMR	0.4	0.4
Mean τ_m (ns)	150.8	338.4
Std τ_m (ns)	22.3	39.0
Max τ_m (ns)	755.7	2077.7
Min τ_m (ns)	2.0 e–1	8.4 e–1
Mean τ_{rms} (ns)	158.8	318.4
Std τ_{rms} (ns)	24.4	46.9
Max τ_{rms} (ns)	635.8	1458.0
Min τ_{rms} (ns)	2.3	3.3

parameters are also given in the row labelled 'Num Prof'. Parameters extracted from small numbers of parameters must be taken as indicative only.

The time-domain measurement results from the Helsinki urban environment were similar to the Sydney environment, however the measurements taken in metropolitan Helsinki exhibited many more prevalent late, large MPCs. This effect was reflected in the much larger values of delay spread and lower CMR. The increased resolvable delay spread also lead to more MPCs being detected. One of the worst examples of such phenomena in the Helsinki metropolitan measurements is given in Figure 6.6. In this measurement, the LOS between transmitter and receiver was obstructed by a wall and the transmitter was significantly higher than the receiver (13 and 5 storeys respectively). The room in which the receiver was kept had windows facing directions away from the transmitter. The resultant measured IR is most likely due to large, strong reflectors in the form of isolated buildings in the area. The reflections from these buildings were able to reach the receiver via paths which represented lower attenuation, for example through windows, than the first received MPC which was obstructed by the room ceiling and walls.

The suburban Helsinki typically has a similar number of resolvable MPCs as the Sydney measurements, however it has a larger CMR and greater values of RMS delay spread. This behaviour may be attributed to the differences in the meaning of 'suburban' residences between Sydney and Helsinki. In Helsinki, the suburban environment consists of widely separated apartment blocks of 2–3 storeys, whereas the Sydney environment consists of closely spaced 'Victorian' terrace houses of uniform height.

For the Sydney measurements, a comparison may be made for the omni-directional and the directional antennas in both urban/suburban environments and for STILL and MOVE measurements. The measurements made with the omni-directional transmit antenna also experienced late, large MPCs. Owing to the shortened resolvable excess

Table 6.6 Channel parameters from Sydney measurements

Model	SYDSUBMO	SYDSUBSO	SYDURBMD	SYDURBMO	SYDURBSD	SYDURBSO
Profiles	2512	4408	7808	10880	13942	9856
Mean path	22.4	18.7	15.4	8.4	12.6	18.4
Std paths	5.1	2.0	1.6	1.1	1.8	1.7
Max paths	73	37	29	28	23	36
Min paths	6	6	1	1	1	2
Mean CMR	0.1	0.9	4.1	168.3	1.9	215.9
Std CMR	0.4	0.6	2.0	135.4	0.4	53.9
Max CMR	6.2	6.8	80.1	935.8	43.4	977.8
Min CMR	0.8	0.5	0.3	0.9	0.4	0.4
Mean τ_m (ns)	124.8	84.8	407.1	70.8	114.7	98.3
Std τ_m (ns)	34.2	17.0	41.3	25.1	11.8	12.3
Max τ_m (ns)	540.5	185.1	691.3	459.9	262.3	1369.3
Min τ_m (ns)	15.6	11.8	–	–	–	–
Mean τ_{rms} (ns)	98.7	79.2	174.9	58.9	100.0	102.3
Std τ_{rms} (ns)	26.3	5.7	19.4	6.1	9.5	11.1
Max τ_{rms} (ns)	438.6	164.0	349.0	177.3	230.4	1645.9
Min τ_{rms} (ns)	37.5	30.5	–	–	–	–

delay of the MOVE measurements, these late MPCs did not appear in the statistics for the MOVE measurements. The STILL measurements will therefore give a better indication of the overall channel parameters. The MOVE type measurements are intended primarily to give an insight into the channel dynamics.

The number of resolvable MPCs for the Sydney Urban STILL measurements are similar for both transmitter antenna types. As may be expected, however, a wider beam of omni-directional antenna leads to more possible paths to the receiver and so does a larger number of MPCs. The relatively low standard deviation for the number of MPCs for both the urban, STILL measurements indicates that this phenomenon was consistent for all receiver positions.

One surprising result from the Sydney urban STILL measurements is that the directional antenna produced a significantly reduced CMR compared to the omni-directional antenna. As mentioned previously, in no scenario was there ever an LOS between transmitter and receiver. The directional antenna measurements always had the beam illuminating the general direction of the receiver location, however the LOS was obstructed by concrete/brick ceilings and walls. The broader beam of the omni-directional antenna allowed more paths between the transmitter and receiver to be utilized. Owing to the different attenuation of paths travelling through walls, windows and various internal fixtures, this 'angle of arrival' diversity allows specific signal elements to have much larger amplitudes than may be expected simply due to conventional path loss analysis. Therefore, in many cases with the omni-directional antenna, a *quasi-line-of-sight* (QLOS) existed in which a strong specular reflector lead to a late arriving MPC which had greater amplitude than that associated with a heavily attenuated first component (NLOS). For the MOVE-type measurements, the limited excess delay does not show these later MPCs and so the CMR statistics do not reflect the same trend.

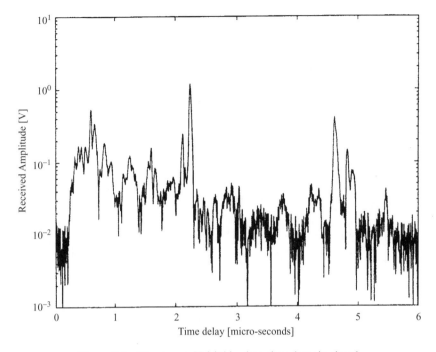

Figure 6.6 Worst-case Helsinki urban time-domain signal

The RMS delay spread statistics show that the arrival of the power of the signal, however, is more evenly matched between the directional and omni-directional transmit antennas. The average RMS delay spread values are approximately 100 ns for both urban, STILL measurements. The maximum values for the omni-directional antenna are significantly greater than that experienced using the directional antenna. This is again due to the presence of very late, large MPCs not found in the directional antenna. This is again due to the presence of very late, large MPCs not found in the directional antenna measurements. The averaged effect of these large, late MPCs does not dominate the delay spread as may be seen from the values for the mean and standard deviation.

Figure 6.7 shows an example of an average (over 256 measurements) IR for the Sydney urban STILL measurements using the omni-directional transmitting antenna for one of the measurement scenarios. Figure 6.8 shows an example of an average (over 256 measurements) IR for the Sydney urban STILL measurements using the directional antenna taken with the same (to within a few centimetres) transmitter/receiver positions. Due to the static nature of the channel, the averaging process serves primarily to reduce the noise floor and so increase the dynamic range of the measurements.

The Sydney suburban measurements show results similar to the urban environment using the omni-directional transmit antenna in terms of number and standard deviation of MPCs. The CMR values are much lower implying that there is in general no dominant MPC which may be found at any delay. This may well be due to the fact that the antenna was just above roof-top height and there were many buildings of similar height near the receiver which was typically on the ground floor of the residences. The idea of more uniform attenuation of all MPCs is brought out in the RMS delay spread statistics. The mean, standard deviation and maximum values are all noticeably lower than the urban

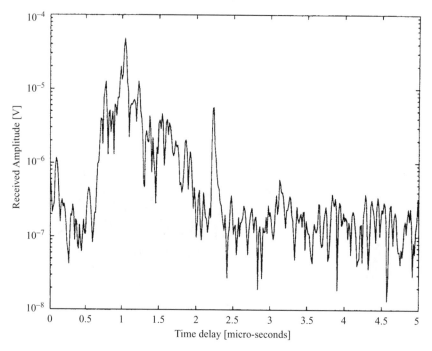

Figure 6.7 Example average time-domain impulse response with omni-directional transmitter in urban environment with no movement

measurements. Figure 6.9 shows an averaged IR for a suburban STILL measurement. A more conventional exponential decay profile is evident from this figure.

6.4.2. Multipath Component Arrivals

One of the features to be examined and reproduced when modelling the environment is the rate at which MPCs arrive and are distributed. For each measured IR, the time after the first component are divided into short, fixed length 'delay windows' and a count is made in each window of the number of MPCs for each IR. It has been noted in several publications [9,12,13] that MPCs tend to arrive in clusters. This is due to the fact that many strong reflectors offer several reflective surfaces or mechanisms for MP arrival. These surfaces are physically closely located so that the path lengths for the MPC are similar leading to closely spaced MPCs at the receiver. The number of MPCs in a window of fixed time length lead to an average arrival rate of MPCs in that window.

For a given propagation model considered, a number of MPC arrival rates are defined. The number of arrival rates to use and the number of delay windows to examine in the model are chosen before the raw IR measurement data is processed. These allowable arrival rates are then used to 'bin' the number of MPCs counted in each arrival window for each measured IR. When the count of each arrival rate for each delay window is complete, the probability of a particular arrival rate occurring in a given delay window is determined.

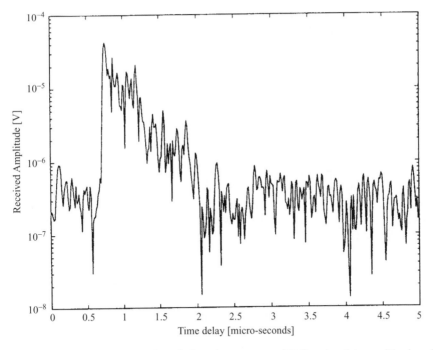

Figure 6.8 Example average time-domain impulse response with directional transmitter in urban environment with no movement

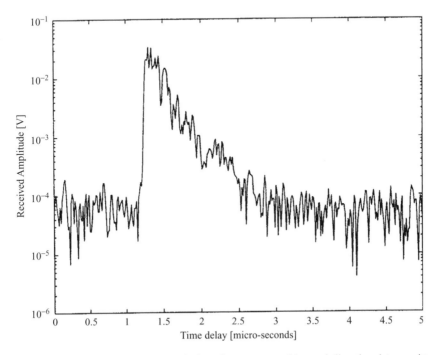

Figure 6.9 Example average time-domain impulse response with omni-directional transmitter in suburban environment with no movement

Figure 6.10 shows the experimentally determined probabilities for each window of length 125 ns (10 samples) receiving 0–5 MPCs for the Sydney Urban, STILL directional measurements. This window size has been chosen based on the observation of the MPC clustering characteristics. As expected, the most likely situation is that the number of MPCs is initially quite high but quickly falls off as the excess delay increases. After some 10 delay windows, the probability of receiving more than 0 MPCs is low.

6.4.3. Multipath Amplitude Variations

Information about amplitude variations is used for the reconstruction of the dynamic behaviour of the channel during a simulation. The usual approach follows is that of D. C. Cox [11] where Doppler information is obtained from the sequence formed by individual MPC amplitude variations. This would then be used to filter AWGN to obtain a series of MPCs with similar statistical characteristics as the original amplitude distribution. For this propagation environment, this is not possible as the Doppler spectrum of STILL measurements had a bandwidth which was very small compared to the resolution of the measurements, so no significant frequencies could be identified. For MOVE measurements, the amplitude variations are well modelled by a Rician distribution. The Doppler spectra associated with these measurements is small in magnitude and has a maximum value well beyond the maximum Doppler shift resolvable by the measurement techniques used and so appears to be flat.

For this reason, the amplitude variations for each MPC are modelled by a Rician distribution. The magnitude of the diffuse component for identified MPC is taken from the variation observed over consecutive measurements (128, 256 or 512) in the samelocation. When the Rice factor calculated from the MPC amplitude variation is plotted

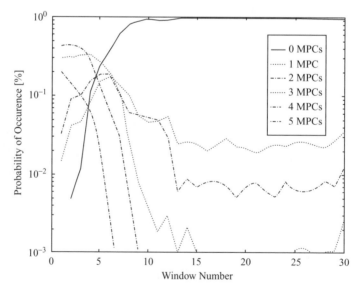

Figure 6.10 Probability of given number of MPCs occurring in each window for Sydney urban, still, directional transmitter model

against the mean MPC amplitude, a log–log relationship is apparent. Figure 6.11 shows the correlation between the measured Rician factor and MPC amplitude for the Helsinki Urban measurement scenario.

6.5. Channel Model

Consideration of the variability of MPCs is one of the most important novel aspects of the channel model being implemented (Figure 6.12). Correct representation of these dynamics are important for the development of a system which will operate over such a WLL channel. The features of the models proposed will be explained with reference tap delay model shown in Figure 6.13 and the measured IR in Figure 6.14. The features discussed are, the number of and location of MPCs, IR power decay profile, amplitude distribution of individual MPCs, amplitude distribution due to shadowing, late arrival of large MPCs. Separate models are specified for each combination of transmitter type, urban or suburban environment and movement classification.

6.5.1. Number of MPCs

When an instance of a model is realized at the start of a simulation, an arrival rate, or number of MPCs to the model, is chosen for each delay window. These arrival rates will be used for the duration of the simulation. The arrival rates to use for any delay window

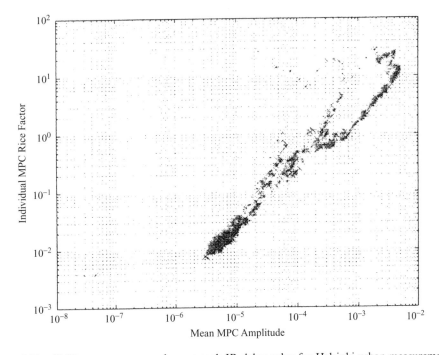

Figure 6.11 CMR versus mean envelope at each IR delay value for Helsinki urban measurements

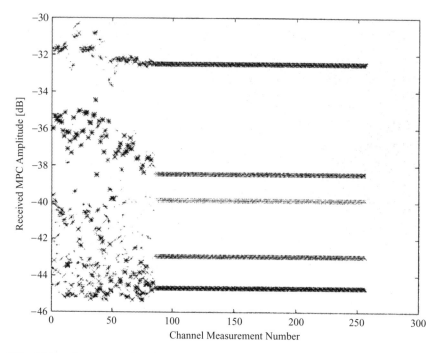

Figure 6.12 Effect of movement immediately near receiver on MPC amplitude variation. Sydney urban MOVE2 measurement, omni-directional transmitter

are chosen in a probabilistic manner from the available rates used in the data process-ing stage. The probability of choosing a given multipath arrival rate will depend on rates chosen for earlier windows and transition probabilities between arrival rates. In other words, the arrival rate for a particular window in a given realization of a model is chosen using a Markov process. The available rates themselves and the prob-abilities of changing rates between consecutive windows are taken from the measured data itself.

At a given delay window, the transition probability matrix for a K arrival rate model is given in the form

$$P_l = \begin{bmatrix} p_{l,1,1} & p_{l,1,2} & \cdots & p_{l,1,K} \\ p_{l,2,1} & p_{l,2,2} & \cdots & p_{l,2,K} \\ \cdots & \cdots & \cdots & \cdots \\ p_{l,K,1} & p_{l,K,2} & \cdots & p_{l,K,K} \end{bmatrix} \tag{6.7}$$

where $p_{l,i,j}$ refers to the probability of the model changing from the current state i to the new state j in window $l + 1$. The matrix of probabilities of selecting a given arrival rate for window l is defined as

$$R_l = [R_{l,1}, R_{l,2}, \ldots, R_{l,k}] \tag{6.8}$$

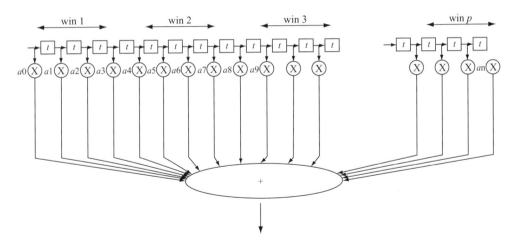

Figure 6.13   Tap delay IR model showing windowing structure

and must satisfy the conditions that

$$R_l P_l = R_l \tag{6.9}$$

$$R_l E = I \tag{6.10}$$

where E is a column matrix whose entries are 1's and I is the identity matrix.

For the description above, each propagation model requires a P_l and R_l matrix for each delay window considered. With a size of 125 ns, approximately 40 delay windows would need to be defined for each model to accurately represent the measured data. Figure 6.13 shows that the probability of a non-zero arrival rate rapidly decreases for later delay windows after the first component. This serves to simplify the model considerably.

Once the arrival rate, or a number of MPCs for a window is chosen, the MPCs are distributed randomly in the window. This configuration remains fixed for the duration of the simulation. In Figure 6.13, a model is shown where a window is defined to be 4 MPCs. In each of these windows, the tap values (a_i) will be non-zero if an MPC is being modelled at that position. The tap value a_0 refers to the first path which may or may not be an LOS.

6.5.2.   IR Power Decay Profile

It has been observed that the measured IRs rarely follow the classical exponential decay profile. This is chiefly due to the different attenuation that each MPCs experiences. As the receiver is always considered to be inside a building in this study, the expected attenuation include a complex series of walls, partitions, windows and objects from the indoor channel, as well as losses associated with the outdoor environment due to reflections, diffraction and free space losses. For this reason each model contains a specification of the mean power and a variance of the received signal at each delay. Based on the examination of the measured data, it is possible to represent the allowable MPC amplitudes at a given delay using a Gaussian distribution of the mean value plus the variance.

Once an amplitude for a given delay is chosen, this value represents the mean amplitude for that MPC for the duration of the simulation. Variations in amplitude of an MPC during a simulation are covered in the next section. From observation of the measured values, the amplitude values at a given delay vary significantly between measurement locations for both STILL and MOVE measurements. This is to be expected as the sources of attenuation are highly location dependent. In Figure 6.14, an IR is shown with an idealized mean amplitude value 'template' from its associated model superimposed. This dashed line represents the value at each delay which will be used as the mean for the determination of the power of an MPC if it appears at that point. Once the amplitude of an MPC is chosen, it remains fixed for the duration of the simulation.

Example templates based on measured values used to generate MPC amplitude levels for a given simulation instance are shown in Figure 6.15. This figure shows the mean measured IR amplitudes, the maximum of all measured IRs and the values generated for this instance of the model.

6.5.3. Amplitude Distribution of Individual MPCs

Once an MPC position and mean amplitude value are chosen, the amplitude of the MPC is allowed to vary over the duration of the simulation. From the observation of the measured IRs, MPC amplitudes at a given delay exhibited distributions which were well

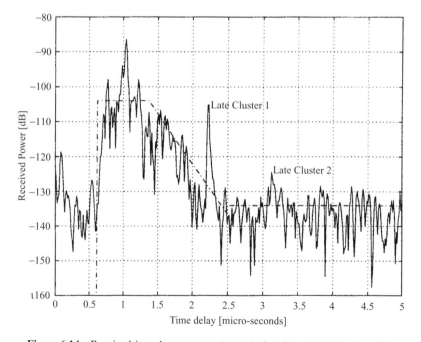

Figure 6.14 Received impulse response (power) showing amplitude template

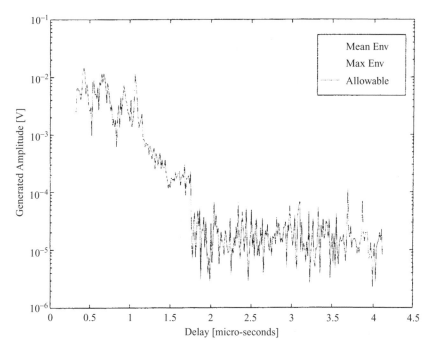

Figure 6.15 Template mean and maximum values used to generate allowable amplitudes a each delay

approximated by Rician PDFs. Each modelled MPC therefore has a complex valued, time-varying diffuse component added to the average amplitude chosen.

As stated in the previous section, in a given measurement position, a strong relationship was identified between the MPC amplitude and the Rice factor used to characterize the MPC amplitude variation. When averaged over all measured IRs in a given environment, this relationship between amplitude and Rice factor is still apparent. The average Rice factors versus MPC amplitudes are therefore used to model amplitude variations for a generated MPC of a given amplitude.

The use of a Rician p.d.f. was applied to MPCs modelled in both MOVE and STILL environments. Unlike for purely indoor channels, the effect of movement in the vicinity of the receiver for the outdoor to indoor channel examined in this paper may still be modelled by a Rician distribution but with increased variation of the MPC amplitudes. Figure 6.12 illustrates the effect of people moving near the receiver on the fading characteristics of the MPCs. This figure shows the MPC with movement (MOVE2) which suddenly stops when the source of the obstruction leaves the room.

The distance between the transmitter and receiver and the corresponding angular dispersion of the incident ray means that the person moving near the receiver may not completely obstruct the arriving MPC (which may be more likely in a purely indoor system). This fact combined with the number of possible arrival angles implies that the level of correlation between successive MPCs amplitudes effected by fading is small. The increase in the number of weak MPCs due to scattering from the obstruction is also less significant than in purely indoor channels. As a result, the fading associated with the movement near the stationary receiver is not modelled as correlated between successive MPCs.

6.5.4.   Late Arrival of Large MPCs/Clusters

It has been observed in previous sections that for outdoor to indoor Urban channel measurements, there is often large, late MPCs which do not fit the simple template model of allowable amplitudes described above. These components are most likely due to distant, strong reflectors and are therefore highly location dependent. It has also been observed that there is typically a cluster of minor MPCs associated with the strong reflection. These smaller components decay exponentially supporting the idea that the strong component is from a strong distant reflected one.

In the urban and metropolitan models, a random number of large reflectors, plus associated minor components, may be selected to model this effect. The maximum and minimum attenuation of these late MPCs may be chosen and the delays specified to be within certain regions of the resulting IR. A more complete specification of the amplitude of the strong reflector is difficult owing to the highly variable nature of attentions of all paths. In Figure 6.14, the location of two such late IR clusters is shown.

6.5.5   Simulation Results

A model which is capable of reproducing all of the channel characteristics described in the sections above has been implemented. This model is capable of running on PC and UNIX platforms and may be used to specify impulse responses for the real-time, radio frequency (RF) channel synthesizer 'PropSim' developed by Elekrobit OY.

An example of a particular channel being selected from a model template is shown in Figure 6.16. Here, three late MPCs are specified and the MPC amplitudes are taken from

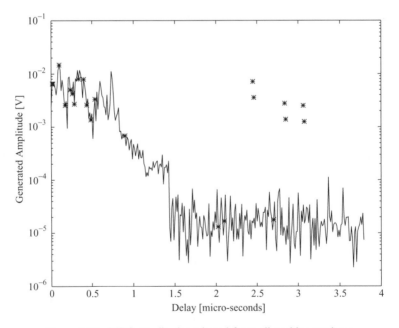

Figure 6.16   MPC amplitudes selected from allowable templates

the amplitude templates shown in Figure 6.15. A series of 50 impulses are shown in Figure 6.17 which are generated from the selected MPC positions and amplitudes, combined with the allowable variances for each delay. These figures clearly show that the main characteristics of the channel have been modelled. The channel parameters of a given instance of the model will be highly dependent on the exact amplitudes and MPC locations selected.

The channel parameters calculated from regenerated instances of each model are given in Tables 6.7 and 6.8 for the Helsinki and Sydney models, respectively. The number of generated IRs used to extract the parameters are also given in the row labelled 'Num Prof'. The values in these tables represent the results of one instance of each model. No late MPCs were modelled as this is the more typical case.

As may be expected from the small, finite number of profiles generated, the range of values for each parameter are considerably smaller than that found in the tables from the measured values. The regenerated profiles show good agreement with the measured values in terms of the number of MPCs, the mean values of CMR factor and delay spread. The values of CMR and delay spread for the generated IRs are well within the 'typical value' range which may be expected from the mean and standard deviation values in Tables 6.5 and 6.6. The variation of CMR and delay spread for a generated set of IRs will not be present as for the measured system as only one example receiver location is generated whereas the values in Tables 6.5 and 6.6 are based on averages over many receiver locations.

The range of values are reasonable and indicate the appropriateness of the approach to the modelling of channel dynamics. One exception, however, is the CMR values for the Sydney, urban, STILL model using the omni-direction antenna. A large number of the measured IRs had very high values (leading to the high standard deviation in Table 6.5). The average profile, however, is significantly more uniform in MPC amplitudes leading to lower CMR values being produced in the regenerated profiles.

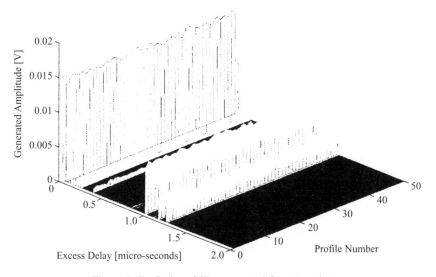

Figure 6.17 Series of IRs generated from templates

Table 6.7 Generated channel parameters from Helsinki models

Name	HELSUBURBMO	HELURBMO
Num Prof	512	512
Mean MPC	22	25
Mean CMR	0.2	0.3
Std CMR	1.0 e–2	1.0 e–2
Max CMR	0.4	0.4
Min CMR	0.1	0.2
Mean τ_m (ns)	298.1	339.6
Std τ_m (ns)	6.6	12.2
Max τ_m (ns)	317.8	379.0
Min τ_m (ns)	278.9	305.7
Mean τ_{rms} (ns)	102.2	242.5
Std τ_{rms} (ns)	5.4	12.4
Max τ_{rms} (ns)	119.1	276.1
Min τ_{rms} (ns)	88.9	209.4

Table 6.8 Generated channel parameters from Sydney models

Model	SYDSUBMO	SYDSUBSO	SYDURBMD	SYDURBMO	SYDURBSD	SYDURBSO
Profiles	512	512	512	512	512	512
Mean paths	17	18	21	7	15	21
Mean CMR	0.1	0.9	0.6	8.3	0.9	0.9
Std CMR	0.4	0.6	0.3	6.4	0.5	0.4
Max CMR	6.2	6.8	2.2	46.8	3.8	3.6
Min CMR	0.8	0.5	0.2	0.4	0.3	0.2
Mean τ_m (ns)	124.8	84.8	467.3	127.1	234.1	652.2
Std τ_m (ns)	34.2	17.0	25.9	22.5	28.2	109.8
Max τ_m (ns)	540.5	185.1	537.7	263.6	310.4	1232.8
Min τ_m (ns)	15.6	11.8	366.8	100.3	150.9	423.5
Mean τ_{rms} (ns)	98.7	79.2	138.5	84.9	135.8	805.8
Std τ_{rms} (ns)	26.3	5.7	22.2	27.8	19.0	167.0
Max τ_{rms} (ns)	438.6	164.0	211.7	173.9	187.2	1319.2
Min τ_{rms} (ns)	37.5	30.5	76.8	29.4	76.9	372.1

As mentioned earlier in the description of the model, once an MPC position is chosen, it remains fixed for the duration of the simulation. This explains the fact that no variance on the numbers of MPCs is given.

References

[1] L. Fernandes, 'Developing a System Concept and Technologies for Mobile Broadband Communications,' *IEEE Personal Commun.—Special Issue*, vol. 2, no. 1, pp. 54–59, 1995.

[2] B. H. Fleury and P. E. Leuthold, 'Radiowave Propagation in Mobile Communications: an Overview of European Research,' *IEEE Commun. Mag.*, pp. 70–81, 1996.

[3] D. Chow, 'Today's Wireless Local Loop Options—Product Survey,' *Mobile Commun. International Mag.*, pp. 47–51, 1996.

[4] W. Mohr, 'Radio Propagation for Local Loop Applications at 2 GHz,' in *Proceedings of the IEEE International Conference on Personal Communications*, Piscataway, NJ, USA, pp. 119–123, 1994.

[5] M. Z. Win, F. Ramirez-Mireles, R. A. Scholtz and M. A. Barnes, 'Ultra-wide Bandwidth (UWB) Signal Propagation for Outdoor Wireless Communications,' in *Proceedings of the IEEE Vehicular Technology Conference*, Pheonix, AZ, USA, May 1997.

[6] R. J. C. Bultitude, P. Melancon, H. Zaghloul, G. Morrison and M. Prokki, 'The Dependence of Indoor Radio Channel Multipath Characteristics on Transmit/Receive Ranges,' *IEEE J. Selected Areas Commun.*, vol. 11, no. 7, pp. 979–990, 1993.

[7] H. Hashemi, 'The Indoor Radio Propagation Channel,' *Proc. IEEE*, vol. 81, no. 7, pp. 943–968, 1993.

[8] F. J. Harris, 'On the Use of Windows for Harmonic Analysis with the Discrete Fourier Transform,' *Proc. IEEE*, vol. 66, no. 1, pp. 943–968, 1978.

[9] E. S. Sousa, V. M. Jovanovic and C. D. Daigneault, 'Delay Spread Measurements for the Digital Cellular Channel in Toronto,' *IEEE Trans. Vehicular Technology*, vol. 43, no. 4, pp. 837–846, 1994.

[10] I. Oppermann, B. White and B. S. Vucetic, 'A Markov Model for Wide-band Fading Channel Simulation in Micro-cellular Systems,' *IEICE Trans. Communications—Special Issue on Personal Communications*, vol. E79-B, no. 9, 1996.

[11] D. C. Cox, 'Delay Doppler Characteristics of Multi-path Propagation at 910 MHz in a Suburban Mobile Radio Environment,' *IEEE Trans. Antennas Propagation*, vol. 40, no. 4, pp. 721–730, 1992.

[12] A. M. Saleh and R. A. Valenzuela, 'A Statistical Model for the Indoor Multipath Propagation,' *IEEE J. Selected Areas Commun*, vol. SAC-5, no. 2, pp. 128–137, 1987.

[13] W. C. Jakes (Editor) *Microwave Mobile Communications Reissue*, IEEE Press, New York, USA, 1994.

[14] J. G. Proakis, *Digital Communications, McGraw-Hill International Editions*, Second edition, McGraw-Hill, New York, 1989.

[15] I. Oppermann, J. Graham and B. S. Vucetic, 'Modelling and Simulation of Indoor Radio Channel at 20 GHz,' in *Proc. GLOBECOM '95*, Singapore, pp. 744–748, 1995.

7

Traffic Considerations in Comparing Access Techniques for WLL

Stefan Mangold, Ingo Forkel, Roger Easo and Bernhard Walke

7.1 Introduction

The focus of interest is the multiple access technology that will be employed in *Wireless Local Loop* (WLL) systems, here referred to as *Fixed Wireless Access* (FWA) networks. The discussion of whether to employ *Code Division Multiple Access* (CDMA) or *Time Division Multiple Access* (TDMA) has gone on for a long time with no result to be expected in the near future. In this chapter, a capacity comparison for FWA based on two access technologies is performed. The TDMA system is analysed with the help of a simulator in chapter by S. V. Krishnamurthy et al. [8]. With this approach it is possible to adjust different scenarios and system parameters in order to find the best capacity utilization of the system. The CDMA system is investigated analytically in this chapter. It is clear that in performing an analytical calculation certain simplifications and assumptions will need to be made.

The structure of this chapter is as follows.

In the next section the FWA network is explained in more detail. Apart from the technological description of this access scheme, the section will also examine the economic viability. The impact of the FWA network in developed and developing countries is investigated and a prognosis is made on the future market possibilities of FWA networks. Following is a brief overview of multiple access schemes.

A derivation for capacity equations of a CDMA system is contained in the following section. An analysis is performed for a single cell scenario offering only one service class. The approach is then adopted for a single radio cell with multiple service classes and finally for a multiple radio cell environment with multiple service classes.

A comparison of the capacity result for TDMA and CDMA is presented at the end of this chapter. It is shown how the expected capacity for both technologies can be estimated.

7.2   Fixed Wireless Access Networks

The accessing of telecommunication services such as telephone, fax and Internet is taken for granted in the developed countries [6,11]. It is therefore surprising to note that the world average teledensity (number of telephone lines per hundred people) is less than 10 %. In fact, almost half of the world's population has never made a phone call. The demand for communication is driven not only by business alliances and exchanges but also through personal relations like friends and relatives that live around the globe. This revolution in communication requirement is abetted by three major forces. Computing power increases while the costs of providing this power are reduced through economies of scale. Secondly the cost of providing transmission of information has fallen by a factor of 10 000 over the last 20 years. Finally the convergence of telecommunications and computing have pushed the merging of segmented industries into a large information industry.

The world information technology market which includes products such as personal computers, mobile phones, and communication has grown by 12.2 % between 1985 and 1995. This is a growth five times faster than the average world *Gross Domestic Product* (GDP) [19].

It is without a doubt established that delivering telecommunication is akin to delivering knowledge. For developing countries delivering knowledge can mean fighting illiteracy and poverty. Therefore, especially these countries need to increase their teledensity. The *International Telecommunications Union* (ITU) recommends that the teledensity of a nation should be at least 20 % so that economic growth is not hampered by the lack of telecommunications.

Wireless access systems provide a suitable method of providing this access to telecommunication services. Currently wireless telephony is experiencing a tremendous growth for the last 10 years with the number of subscribers globally estimated at 55 million people until mid 1995 [7]. Most of this usage is for mobile communications.

The prediction of which access technique will have the greatest impact must be based on the current tariff ideology. The tariff structure of telephone calls does not provide resemblance to the real cost involved. The highest costs are incurred in the local loop and is proportional to the distance of the subscriber to the distribution point. This would mean that a call coming from a rural area should be more expensive than an urban call. Further, an international call should be only nominally more expensive than a local call. These facts are not reflected in the current tariffs. This is partly due to the monopolistic history of most *Public Telephony Operators* (PTO). Being regulated by the national governments, the PTO had the obligation of offering each citizen a connection at a universal price.

The advent of Internet telephony will break the current tariff structure. Allowing Internet users to perform international calls at the local call rate. This application is a true reflection of the actual situation. Since operators inflate the cost of international calls to reduce the loss made on subsidising the local loop, popular Internet telephony will make its impact. The result will be a restructuring of the tariff system. This so-called voice over *Internet Protocol* (IP) is not expected to last very long as the Internet will be flooded with voice and data movement.

The economic impact of FWA networks will largely depend upon the success of *Digital Subscriber Line* (xDSL) technologies. These so-called 'killer' technologies have the possibility of rendering all other access techniques obsolete. However, the penetration of xDSL is questionable as some figures state that only 30 % of all telephone lines can be utilized for xDSL. Another problem of xDSL is that the lines are owned by the PTO. There is a certain amount of control which a private operator must relinquish when renting a line from the PTO. It might be that xDSL will need another five years for a breakthrough in the local loop. However, the success of xDSL will be crucial for the existence of other access technologies.

Despite the generous forecasts that were made for FWA networks, some predictions were 170 million subscribers by the year 2000, the impact of this technology has been slow. For 1998, the subscriber count is at best a few million (some say just 1 million).

Companies offering FWA networks in the market have even seen considerable dropping of share value. This is a surprising since wireless access does have a considerable cost advantage over all the other technologies [17]. However, the introduction of FWA is very expensive if a wired solution is present. Further, the operation of FWA networks generally require the acquirement of two licenses, one enabling the offer of telecommunication services and the other the use of the radio spectrum. The allocation of radio spectrum is also a problem. To be able to offer high transmission bit rates, sufficient bandwidth must be allocated. In some cases this allocation has been too low.

Another deciding factor apart from cost will be the subscriber's demand for high-bandwidth services. Test carried out with *Video on Demand* (VoD) and home shopping do not reflect heavy user interest. This is different, however, for teleworkers and business users who need to work with the company network at comparable *Local Area Network* (LAN) schemes.

In summary, it can be said that FWA networks will be a very viable technology for developing countries and Eastern Europe. The higher risk is clearly bound with the deployment in the developed countries.

7.3 Multiple Access Technologies

Presented in this section is a brief description of the two major access technologies for wireless networks. The communication medium for a radio system is a commonly shared radio channel. Considering the uplink, the link from the *Radio Network Terminals* (RNT) to the *Radio Base Station* (RBS) the system can be classified as a *MultiPoint-to-Point* (MPP) system. With multiple access technology it is possible for several users to send their signals over the radio channel which are then ultimately detected at a corresponding receiver.

For the sake of completeness the *Frequency Division Multiple Access* (FDMA) method should be mentioned but is not explained in more detail. For a general overview it can be referred to B. Walke [16].

7.3.1 Time Division Multiple Access

With TDMA the radio resource is divided in the time domain into time slots. The time slots are assigned to users either in a cyclic fashion or upon demand. Within this time slot an exclusive user is able to transmit across the medium. To avoid collisions the system must be synchronized and additionally a guard time is inserted between slots. No other conversations can access an occupied TDMA channel until the channel is vacated. TDMA is a software intensive protocol so the gathering of results is possible by means of simulations.

Figure 7.1 illustrates the basic principle of TDMA with the alternating transmission and guard periods.

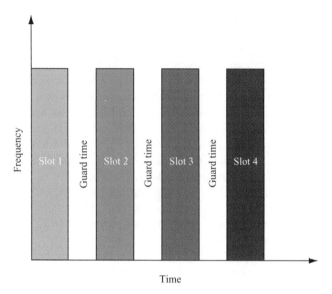

Figure 7.1   TDMA (Walke, 1999)

TDMA is a common multiple access technique employed in digital cellular systems. Its standards include North American Digital Cellular, *Global System for Mobile Communications* (GSM), and *Personal Digital Cellular* (PDC).

7.3.2   Code Division Multiple Access

CDMA is a form of spread-spectrum, an advanced digital wireless transmission technique. Instead of using frequencies or time slots, as do traditional technologies, it uses mathematical codes to transmit and distinguish between multiple wireless conversations [10]. Its bandwidth is much wider than that required for simple point-to-point communications at the same data rate because it uses noise-like carrier waves to spread the information contained in a signal of interest over a much greater bandwidth. However, because the conversations taking place are distinguished by digital codes, many users can share the same bandwidth simultaneously, as seen in Figure 7.2.

Although not shown, it is possible for a user to use more than one code, as is foreseen for third-generation mobile systems. The advanced methods used in commercial CDMA technology improve capacity, coverage and voice quality, leading to a new generation of wireless networks.

7.3.3   Interference in Multiple Access Systems

A multiple access scheme must warrant that a user can access the radio channel without causing interference to the other users. If interference is caused, it is then known as *Multiple Access Interference* (MAI), interference caused by the multiple accession to the radio channel. In the presence of MAI the data symbols of the different users interfere with each other.

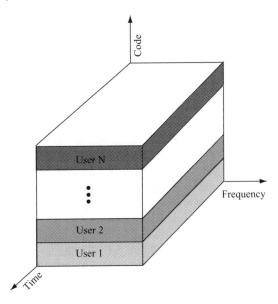

Figure 7.2   CDMA (Rappaport, 1996)

If there is multipath propagation on the channel, then the symbols in the signal of a single user cause interference upon each other, leading to *Inter-Symbol Interference* (ISI). ISI takes place if the symbol duration is less than the time dispersion on the channel, a phenomenon which can take place if the transmission bit rate is very high. Both MAI and ISI can be grouped together and classified as intra-cell interference, the interference present in a radio cell. A radio cell in a multicellular environment additionally experiences interference caused by the transmitting stations in neighbouring radio cells. This interference is known as inter-cell interference.

7.4   CDMA Capacity Analysis

Presented here is an analytical method to determine the capacity of a multiclass multicellular spread sequence (CDMA) systems based on an approach by S. J. Lee et al. [9]. The basis of the method assumes an a-priori E_b/I_0 level which must be maintained to assure a satisfactory performance with respect to the *Bit Error Ratio* (BER) for a desired service class. Capacity is defined here as the number of simultaneous connections that can be admitted into the system for a particular service class so that the quality constraint can still be guaranteed. The capacity analysis is carried out for the reverse link (RNT to RBS uplink) since this link is considered to be critical for a CDMA system [8].

7.4.1   CDMA Traffic Model

The aim of a broadband FWA network is to carry different types of service classes, each requiring a different service bit rate. A survey conducted for integrated services on wireless multiple access networks has come up with a possible service performance for these networks [12].

Bit-Energy to Interference Spectral Power　The bit-energy to interference spectral power denoted here as $\gamma = E_b/I_0$ is the constraining factor for a CDMA system when allocating capacity to a new connection. The term is mainly dependent on the maximum BER the service can sustain and the modulation type selected for the transmission.

Spreading Gain　The spreading gain G (equalling the spreading factor in a CDMA system) depends on the service bit rate, the transmission bandwidth and the multirate transmission technology. For the *Single-Code* (SC) technology there are different values of G since different bit rates are realized by different spreading of the data sequence. Whereas for *MultiCode* (MC) technology there is only one spreading gain equal for all codes used, but a number of codes can be multiplexed in order to offer the required transmission bit rate.

Considering the service bit rates from Table 7.1 and the transmission bandwidth $W = 112$ Mbit/s, a certain spreading gain G for the services could be assigned as proposed in Table 7.2. The base transmission bit rate R_b for MC-CDMA was chosen to be the lowest service bit rate of the system, the bit rate for the voice calls. The spreading gain is the quotient of transmission bandwidth to service bit rate.

7.4.2　Single-Class Services

This is the most common type of capacity analysis for a *Direct Sequence CDMA* (DS-CDMA) system. Generally the service class under scrutiny are voice calls with a service bit rate of 32 kbit/s. The resulting capacity equation derived here is of little importance for a FWA network desired to work on a broadband system.

Table 7.1　Service classes for FWA networks

Service	Maximum BER	Delay	Bit rate	required γ
Class 1 (voice)	10^{-3}	Sensitive	32 kbit/s	6.8 dB
Class 2 (Packet Data)	10^{-4}	Insensitive	64 kbit/s	7.0 dB
Class 3 (video)	10^{-5}	Sensitive	128 kbit/s	9.5 dB

Table 7.2　Spreading gains for different service classes

Service	SC-CDMA	MC-CDMA	Bit rate
Class 1 (voice)	$G = 3500$	$G = 3500$; 1 Code	32 kbit/s
Class 2 (Packet Data)	$G = 1750$	$G = 3500$; 2 Code	64 kbit/s
Class 3 (video)	$G = 875$	$G = 3500$; 4 Code	128 kbit/s

However, the approach and the trail of thought will be the same one for the pursuit of capacity for a multiple service class system.

7.4.2.1 Single-Cell and Single-Class Capacity

The error rate of digital transmission systems only depends on the signal-to-noise ratio. Respectively the *Carrier to Interference Ratio* (C/I) expressed by

$$\frac{C}{I_{\text{intra}}} = \frac{S}{N} \tag{7.1}$$

where S portrays the sending power reception level of a user signal at a receiver and I_{intra} is the total interference power experienced within a single cell (intra-cell interference).

Assume it is possible to construct a source where the sending signal spectrum is constant between $-W/2 \leq f \leq +W/2$ and disappears outside this interval. Let E_b be the energy per bit of this signal and the bit rate be $R = 1/T$. The sending power is now equal to $E_b R$. The term N_0 in Equation (7.2) is the power density of the noise power N resulting from the effects of thermal noise and spurious interference in the bandwidth. It can be written that

$$E_b = \frac{S}{R} \quad \text{and} \quad N_0 = \frac{N}{W} \tag{7.2}$$

Now consider a system with n sources, each possessing the before described characteristic. The ith receiver correlates the received signal with all the other $n-1$ signals. Assuming that the sending signal of all the other sources are uncorrelated, then the ith receiver regards the other signals as uncorrelated white noise sources. Further, it is assumed that the received power level of the different sources are all equal at the site of the receiver (perfectly power controlled). This yields

$$\frac{E_b}{I_{\text{intra}}} = \frac{\frac{S}{R}}{(n-1)\frac{S}{W}} = \frac{\frac{W}{R}}{n-1} \tag{7.3}$$

This equation can be modified to include noise effects [4]. These effects are contained in the term N_0 found in Equation (7.2)

$$\frac{E_b}{I_{\text{intra}} + N_0} = \frac{\frac{S}{R}}{(n-1)\frac{S}{W} + N_0} \tag{7.4}$$

Extending Equation (7.4) with the term W/S and using the identity for N_0 from Equation (7.2)

$$\frac{E_b}{I_{\text{intra}} + N_0} = \frac{\frac{W}{R}}{(n-1) + \frac{N}{S}} \tag{7.5}$$

Cancelling the denominator, Equation (7.5) can be rewritten as

$$\frac{E_b}{I_0} = \frac{E_b}{I_{\text{intra}} + N_0} = \frac{1}{(n-1)/\frac{3}{2}G + \frac{N_0}{E_b}} \tag{7.6}$$

where I_0 is referred to as the interference power spectral density and G is the spreading gain defined in the other place. The coefficient $3/2$ results from the rectangular chip form of the spreading code [3].

Using the definition $G = R^{\text{chip}}/R$ whereby R^{chip} is the chip rate of the spreading sequence and modifying Equation (7.6) to remove the term E_b the quality constraint for a service finally becomes

$$\frac{E_b}{I_0} = \frac{\frac{S}{R}}{(n-1)S/\frac{3}{2}R^{\text{chip}} + N_0} \tag{7.7}$$

Introducing the term γ for the required bit energy to interference power spectral density ratio E_b/I_0, the constraint for an acceptable connection (considering the BER as a connection admission criterion) is

$$\frac{\frac{S}{R}}{(n-1)S/\frac{3}{2}R^{\text{chip}} + N_0} \geq \gamma \tag{7.8}$$

Hence, the number of simultaneous accepted connections (also called capacity) in a single cell offering only one service is equal to

$$n \leq \frac{\frac{3}{2}G + \gamma}{\gamma} - \frac{\frac{3}{2}R^{\text{chip}}N_0}{S} \tag{7.9}$$

7.4.3   MultiClass Services

The system is now be extended to include the transmission of different service classes. These services can be voice calls, Internet services or data transmission for example.

In the system being analysed there are up to K service classes, each service class having an information bit rate R_k. For single-code transmission this bit rate is an integer multiple of the line bit rate R. In the case of MC transmission, the high information bit rate of a class k connection is defined by $R_k = c_k R$. The term c_k denotes the number of codes needed for transmitting a class k connection [2].

Further, it is assumed that there are n_k connections in each service class k. The connection is linked to the RBS with the least path loss.

Using a similar line of thought as in Equation (7.7) the bit energy to interference power spectral density ratio for the ith connection is modelled as

$$\frac{E_b}{I_0} = \frac{E_b}{I_{\text{intra}} + N_0} \tag{7.10}$$

The value E_b/I_0 is the constraint value for the connection admission and is determined by the modulation technique of the system and the BER which must be guaranteed for the connection.

In the equations below S_i denotes the received level of the signal power of the connection to be accepted. Depending on the transmission scheme, R_i is the ith terminal's service bit rate for the SC system whereas R is the line rate of the MC system. The intra-cell interference is no longer based on the uncorrelated disturber signals but rather on the interference caused by the different connections with their corresponding received power levels.

$$\frac{E_b}{I_0}\bigg|_i^{SC} = \frac{\frac{S_i}{R_i}}{\left(\sum_{k=1}^{K} n_k S_k - S_i\right) \bigg/ \frac{3}{2} R^{\text{chip}} + N_0} \qquad (7.11)$$

$$\frac{E_b}{I_0}\bigg|_i^{MC} = \frac{\frac{S_i}{R}}{\left(\sum_{k=1}^{K} c_k n_k S_k - c_i S_i\right) \bigg/ \frac{3}{2} R^{\text{chip}} + N_0} \qquad (7.12)$$

It is apparent that c_k in Equation (7.12) is the term for the number of codes necessary to service one of the n_k connections of service class k for an MC transmission scheme. Similarly c_i is the number of codes needed for the ith connection under consideration for the analysis. The I_{intra} term resulting from the existing connections is also known as the *Co-Channel Interference* (CCI) of the cell.

$$\frac{E_b}{I_0} = \frac{E_b}{\text{CCI} + N_0} \qquad (7.13)$$

This basic interference limited model needs to be modified to include the effects of the channel in which the spreading sequence is propagated. The behaviour of the channel is modelled as a *Wide Sense Stationary Uncorrelated Scattered* (WSSUS) channel. The multipath propagation induced in this channel leads to interference between the code sequence symbols, hence known as ISI. The I_{intra} is now the sum of both, ISI and CCI

$$\frac{E_b}{I_0} = \frac{E_b}{\text{CCI} + \text{ISI} + N_0} \qquad (7.14)$$

Equations (7.11) and (7.12) are extended to consider ISI. This type of interference is contained in the terms $F(G_i)$ or $F(G)$ [9]. Additionally, a transmission coefficient v^2 describes the ratio of the primary received signal to the multipath signal of the considered interfering links and acknowledge the effect of multipath propagation for these signal components as well.

$$\frac{E_b}{I_0}\bigg|_i^{SC} = \frac{\frac{S_i}{R_i}}{\left(\sum_{k=1}^{K} n_k S_k - S_i\right) \cdot \frac{1 + 2v^2}{\frac{3}{2} R^{\text{chip}}} + \frac{S_i}{R_i} F(G_i) + N_0} \qquad (7.15)$$

$$\frac{E_b}{I_0}\bigg|_i^{MC} = \frac{\frac{S_i}{R}}{\left(\sum_{k=1}^{K} c_k n_k S_k - c_i S_i\right) \cdot \frac{1 + 2v^2}{\frac{3}{2} R^{\text{chip}}} + \frac{S_i}{R} F(G) + N_0} \qquad (7.16)$$

These equations are the basic frame work for the more complex investigation which will follow.

7.4.3.1 Single-Cell and MultiClass Services Capacity

The aim of the capacity analysis is to determine the number of connections that can be simultaneously admitted into the transmission system. The basic parameter for the connection admission is the E_b/I_0 value which is inherently determined by the service quality, e.g. the BER required.

Recalling the identity $E_b/I_0 = \gamma$ and Equation (7.8), the quality constraint condition for a connection i of a particular service class is

$$\frac{\frac{S_i}{R_i}}{\left(\sum_{k=1}^{K} n_k S_k - S_i\right) \cdot \frac{1 + 2v^2}{\frac{3}{2}R^{\text{chip}}} + \frac{S_i}{R_i}F(G_i) + N_0} \geq \gamma_i\Big|^{\text{SC}} \tag{7.17}$$

$$\frac{\frac{S_i}{R}}{\left(\sum_{k=1}^{K} c_k n_k S_k - c_i S_i\right) \cdot \frac{1 + 2v^2}{\frac{3}{2}R^{\text{chip}}} + \frac{S_i}{R}F(G) + N_0} \geq \gamma_i\Big|^{\text{MC}} \tag{7.18}$$

Following simple algebra, these equations can be rewritten as

$$\frac{1 + 2v^2 + \frac{3}{2}G_i\left(\frac{1}{\gamma_i} - F(G_i)\right)}{1 + 2v^2} S_i \geq \sum_{k=1}^{K} n_k S_k + \frac{\frac{3}{2}N_0 R^{\text{chip}}}{1 + 2v^2}\Big|^{\text{SC}} \tag{7.19}$$

$$\frac{(1 + 2v^2)c_i + \frac{3}{2}G\left(\frac{1}{\gamma_i} - F(G)\right)}{1 + 2v^2} S_i \geq \sum_{k=1}^{K} c_k n_k S_k + \frac{\frac{3}{2}N_0 R^{\text{chip}}}{1 + 2v^2}\Big|^{\text{MC}} \tag{7.20}$$

Equations of this form obey the following proposition:

$$a_i S_i \geq \sum_{k=1}^{K} n_k S_k + b \quad \text{exists if and only if} \quad \sum_{k=1}^{K} \frac{1}{1 - \varepsilon} \cdot \frac{n_k}{a_k} \leq 1 \tag{7.21}$$

when

$$\varepsilon = \max_{i \in \{1,\dots,K\}} \left\{\frac{b}{a_i S_i}\right\} \tag{7.22}$$

Applying this proposition for the SC calculations, Equation (7.19) yields

$$\sum_{i=1}^{K} \frac{1}{1 - \varepsilon} \cdot \frac{1 + 2v^2}{1 + 2v^2 + \frac{3}{2}G_i\left(\frac{1}{\gamma_i} - F(G_i)\right)} n_i \leq 1\Big|^{\text{SC}} \tag{7.23}$$

where

$$\varepsilon = \max_{i \in \{1,...,K\}} \left\{ \frac{1}{1 + 2v^2 + \frac{3}{2} G_i \left(\frac{1}{\gamma_i} - F(G_i) \right)} \cdot \frac{\frac{3}{2} N_0 R^{\mathrm{chip}}}{S_i} \right\} \Bigg|^{\mathrm{SC}} \tag{7.24}$$

Similarly, Equation (7.20) for the MC transmission will be transferred to

$$\sum_{i=1}^{K} \frac{1}{1 - \varepsilon} \cdot \frac{(1 + 2v^2) c_i}{(1 + 2v^2) c_i + \frac{3}{2} G \left(\frac{1}{\gamma_i} - F(G) \right)} n_i \le 1 \Bigg|^{\mathrm{MC}} \tag{7.25}$$

with its associated

$$\varepsilon = \max_{i \in \{1,...,K\}} \left\{ \frac{1}{(1 + 2v^2) c_i + \frac{3}{2} G \left(\frac{1}{\gamma_i} - F(G) \right)} \cdot \frac{\frac{3}{2} N_0 R^{\mathrm{chip}}}{S_i} \right\} \Bigg|^{\mathrm{MC}} \tag{7.26}$$

As can be seen, Equations (7.23) and (7.25) are in the form

$$\sum_{k=1}^{K} \alpha_k n_k \le 1 \tag{7.27}$$

Recalling that n_k is the number of existing connections of the service class k in the system, the term α_k becomes the bandwidth allocated to each connection of this service class. More correctly, α_k is the bandwidth allocated to a single connection of service class k with $(k = 1, \ldots, K)$ normalized to the entire bandwidth W of the system. The inverse of this term denotes the number of connections which can be allocated in the entire bandwidth of the system for a predetermined quality (E_b/I_0) of the connection. Hence, it can be written as

$$\alpha_k = \frac{W_k}{W} \tag{7.28}$$

where W_k is the bandwidth of the respective single connection of service class k. The system capacity with respect to connections that can be established is then derived from the inverse of α_k like that the total number of connections of service class k which can be carried by the system is $1/\alpha_k$.

Comparing Equation (7.27) with Equation (7.23) the normalized bandwidth for SC transmission systems is

$$\alpha_i \big|^{\mathrm{SC}} = \frac{1}{1 - \varepsilon} \cdot \frac{1 + 2v^2}{1 + 2v^2 + \frac{3}{2} G_i \left(\frac{1}{\gamma_i} - F(G_i) \right)} \tag{7.29}$$

The normalized bandwidth for MC transmission systems is calculated from Equations (7.27) and (7.25)

$$\alpha_i\big|^{MC} = \frac{1}{1-\varepsilon} \cdot \frac{(1+2v^2)c_i}{(1+2v^2)c_i + \frac{3}{2}G\left(\frac{1}{\gamma_i} - F(G)\right)} \tag{7.30}$$

7.4.3.2 MultiCells and MultiClass Services Capacity

In order to analyse the capacity of a CDMA system for a multicellular layout the interference at the radio cell site caused by the neighbouring radio cells must be calculated. Until now only ISI and CCI resulting from the connections established in one cell has been considered. Returning to the basic quality constraint model, Equation (7.10) is now modified to include the interference from adjacent radio cells, also known as intercell interference I_{inter}.

$$\frac{E_b}{I_0} = \frac{E_b}{I_{intra} + I_{inter} + N_0} \tag{7.31}$$

Similarly to the approach for the single cell scenario the primary equations for the capacity calculation in a multicellular system are

$$\frac{\frac{S_i}{R_i}}{\frac{\left(\sum_{k=1}^{K} n_k S_k - S_i\right)(1+2v^2) + I_{inter}}{\frac{3}{2}R^{chip}} + \frac{S_i}{R_i}F(G_i) + N_0} \geq \gamma_i\big|^{SC} \tag{7.32}$$

for the SC transmission and for MC transmission

$$\frac{\frac{S_i}{R}}{\frac{\left(\sum_{k=1}^{K} c_k n_k S_k - c_i S_i\right)(1+2v^2) + I_{inter}}{\frac{3}{2}R^{chip}} + \frac{S_i}{R}F(G) + N_0} \geq \gamma_i\big|^{MC} \tag{7.33}$$

These equations can also be rewritten in order to reach equations in a form similar to Equations (7.19) and (7.20) in the form

$$\frac{1 + 2v^2 + \frac{3}{2}G_i\left(\frac{1}{\gamma_i} - F(G_i)\right)}{1 + 2v^2}S_i \geq \sum_{k=1}^{K} n_k S_k + \frac{I_{inter} + \frac{3}{2}N_0 R^{chip}}{1 + 2v^2}\bigg|^{SC} \tag{7.34}$$

$$\frac{(1 + 2v^2)c_i + \frac{3}{2}G\left(\frac{1}{\gamma_i} - F(G)\right)}{1 + 2v^2}S_i \geq \sum_{k=1}^{K} c_k n_k S_k + \frac{I_{inter} + \frac{3}{2}N_0 R^{chip}}{1 + 2v^2}\bigg|^{MC} \tag{7.35}$$

Applying the proposition from Equation (7.21), leads to the following expression for SC systems, from which the normalized bandwidth can be inferred

$$\sum_{i=1}^{K} \frac{1}{1-\zeta} \cdot \frac{1+2v^2}{1+2v^2+\frac{3}{2}G_i\left(\frac{1}{\gamma_i}-F(G_i)\right)} n_i \leq 1 \Bigg|^{SC} \tag{7.36}$$

where ζ plays the role of the previous term ε

$$\zeta = \max_{i\in\{1,\dots,K\}} \left\{ \frac{1}{1+2v^2+\frac{3}{2}G_i\left(\frac{1}{\gamma_i}-F(G_i)\right)} \cdot \frac{I_{inter}+\frac{3}{2}N_0 R^{chip}}{S_i} \right\}\Bigg|^{SC} \tag{7.37}$$

Likewise Equation (7.35) for MC systems in a multicellular environment can be transferred to the form

$$\sum_{i=1}^{K} \frac{1}{1-\zeta} \cdot \frac{(1+2v^2)c_i}{(1+2v^2)c_i+\frac{3}{2}G\left(\frac{1}{\gamma_i}-F(G)\right)} n_i \leq 1 \Bigg|^{MC} \tag{38}$$

with

$$\zeta = \max_{i\in\{1,\dots,K\}} \left\{ \frac{1}{(1+2v^2)c_i+\frac{3}{2}G\left(\frac{1}{\gamma_i}-F(G)\right)} \cdot \frac{I_{inter}+\frac{3}{2}N_0 R^{chip}}{S_i} \right\}\Bigg|^{MC} \tag{7.39}$$

These capacity equations can now be used to determine the normalized bandwidth for a single class k connection. Looking at Equation (7.36) which is in the form of Equation (7.27) the bandwidth necessary for a SC connection is equal to

$$\alpha_i\big|^{SC} = \frac{1}{1-\zeta} \cdot \frac{1+2v^2}{1+2v^2+\frac{3}{2}G_i\left(\frac{1}{\gamma_i}-F(G_i)\right)} \cdot \tag{7.40}$$

Correspondingly, comparing Equations (7.38) and (7.27) for MC systems

$$\alpha_i\big|^{MC} = \frac{1}{1-\zeta} \cdot \frac{(1+2v^2)c_i}{(1+2v^2)c_i+\frac{3}{2}G\left(\frac{1}{\gamma_i}-F(G)\right)} \tag{7.41}$$

It is now possible to calculate the system capacity of a CDMA system servicing multiple service classes in a multicellular environment. In order to be able to perform the calculations for system capacity the following system parameters must be known.

Parameter	Description
W	Transmission bandwidth
S_i	received signal power
R_i	service bit-rate
E_b/I_0	bit-energy to interference spectral power density

7.4.4 Analytical Results of CDMA in FWA Networks

Using the information gathered in the sections before, the actual capacity calculation for DS-CDMA in a FWA network can now be carried out.

7.4.4.1 Ideal CDMA System

First a CDMA system with ideal properties is analysed. Ideal in this case means the absence of inter-cell interference and ISI. This means that the multipath propagation coefficient v^2 and I_{intra} are equal to 0. Remaining is only the intra-cell interference I_{intra}. For the sake of completeness it is mentioned that receiver technologies based on joint detection might eliminate this kind of interference when the number of codes used in the cell is sufficiently low. However, in this study such improved reception methods are not considered. For SC transmission with the presence of I_{intra} in this case Equation (7.40) is simplified to

$$\alpha_i\big|^{\text{SC}} = \frac{1}{1-\zeta} \cdot \frac{1}{1+\frac{3}{2}G_i\frac{1}{\gamma_i}} \tag{7.42}$$

with

$$\zeta = \max_{i\in\{1,\ldots,K\}} \left\{ \frac{1}{1+\frac{3}{2}G_i\frac{1}{\gamma_i}} \cdot \frac{\frac{3}{2}N_0 R^{\text{chip}}}{S_i} \right\}\Bigg|^{\text{SC}} \tag{7.43}$$

where ζ approximately takes the value 1. A similar function can be developed from Equation (7.41) for the MC scheme. The number of connections that can be carried is calculated through the inverse of the normalized bandwidth α_i. Table 7.4 gives an overview of the number of connections that can be held in such an ideal system.

For class 1 services both, SC and MC offer the same number of connections. This is because for this service both schemes have the same spreading gain. While for SC the spreading gain decreases as the service bit rate increases, MC still offers a high spreading gain thereby being able to allocate slightly more calls. Nevertheless, the MC transmission scheme looses performance by the increased number of codes used for transmission and with that higher interference.

Table 7.4 Maximum number of possible connections for different service classes in an ideal CDMA system

Service	SC-Connections	MC-Connections
Class 1	1082	1082
Class 2	347	358
Class 3	124	140

Table 7.5 Maximum number of possible connections for different service classes in a CDMA system with inter-cell interference

Service	SC-Connections	MC-Connections
class 1	1079	1079
class 2	344	356
class 3	124	139

7.4.4.2 CDMA System with Inter-Cell Interference

The effect of inter-cell interference on a CDMA system can be seen if this characteristic is considered in the ideal system described before. Since there is no ISI, the equations need to be computed for the case that there is no multipath propagation, therefore v^2 is also equal to 0. Table 7.5 shows the connection parameters of the different service classes if the omni-directional scenario is analysed.

Similar results are achieved for the analysis performed for the scenarios with directional antennas for the RNTs and sectored RBSs. A CDMA system with sectoring shows a maximum increase of 3 connections which can be additionally carried.

7.4.4.3 CDMA System with Inter-Symbol Interference

The CDMA system is degraded if ISI is introduced into the system. However, the Rake receiver technology which benefits from multipath propagation is left out of consideration here. For connections based on an SC technology, Figures 7.3, 7.4 and 7.5 describe the capacity result for class 1, class 2, and class 3 services, respectively.

Shown in the figures is the amount of bandwidth a single connection requires for a certain channel quality and a transmission quality based on the multipath characteristic of the Rice fading channel. The axis denoted with $F(G)$ takes the value 0 for FWA networks according to the explanation for the ISI earlier in this chapter. However, to be fair a certain degrading quality of the channel should be considered. Using Equation (7.54) it can be seen that $F(G)$ or $F(G_i)$ is not allowed to be larger than the inverse of the bit energy to spectral interference density ratio. To recapture, $F(G)$ models the ISI between two adjacent bits dependent on the spreading gain G. If no ISI takes place, this function is 0. The transmission coefficient v^2 describes the ratio of the primary signal to the multipath signal. At $F(G)$ equalling 0, there is no multipath propagation and

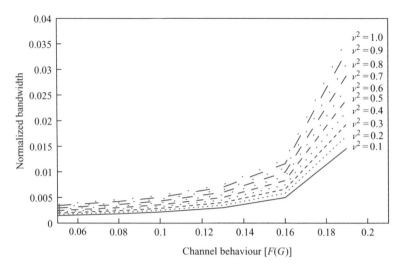

Figure 7.3   32 kbit/s (voice) connection capacity

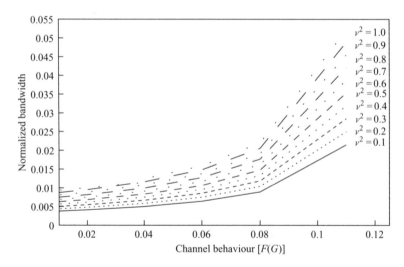

Figure 7.4   64 kbit/s (packet data) connection capacity

hence, v^2 is also 0. The capacity degrades as both $F(G)$ and v^2 rise. In other words, the worse a channel gets, the more bandwidth is required to offer an obligatory E_b/I_0 value.

The class 1 service shows the lowest bandwidth utilization in the CDMA system. This is due to the relatively shallow service bit rate and the low quality constraint required for this class. The range of $F(G)$ here lies in the interval $[0 . . 0.2]$ the limits being determined by the quality constraint. The bandwidth utilization of this service is 0.0009 % making a total of 1079 simultaneous voice connections in one radio cell possible. A radio channel of the worst quality can only offer 27 connections.

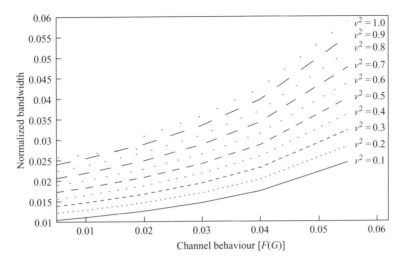

Figure 7.5 128 kbit/s (video) connection capacity

The bandwidth utilization increases with higher service bit rates and more stringent quality demands. This can be seen in Figure 7.4. There is a rise in the bandwidth requirement of the class 2 connections. Now the FWA channel can handle 344 connections, each requiring 0.3 % of the total bandwidth on offer. Due to the effects of multipath propagation causing an increased ISI and I_{inter} in the radio cell, only 19 class 2 connections are possible in such a worst case scenario.

The class 3 video connections show the highest bandwidth utilization. Only 124 simultaneous connections can be offered in one radio cell. The maximum number of connections under the influence of a channel with severe multipath propagation is 17.

The capacity calculations are carried out for both, SC and MC transmission technologies. The trend for MC transmissions is similar to the one for SC with the exception that the utilized bandwidths for service classes 2 and 3 are slightly less. Obviously the differences can only be seen for these services since both technologies have the same spreading gain and number of used codes for the class 1 service. It can be concluded that MC transmission with a high spreading gain would offer more capacity than an SC transmission with variable spreading gains dependent on the bit rates of the services.

Moreover, MC transmission offers more flexibility for time-varying traffic in realistic scenarios.

Interesting is the number of connections that can be held of each service in a realistic CDMA system. In [9] the following values collected in Table 7.6 are presented for the channel parameters v^2 and $F(G)$.

Based on these values it is now possible to present the different number of connections for the three service classes in systems of different qualities.

In a WLL performance study carried out by Deutsche Telekom AG, a maximum of 64 users with a bit rate of 32 kbit/s could be served with a BER less than 10^{-3} [15]. The DS-CDMA system under study had a bandwidth of 3.5 MHz at a carrier frequency of 3.5 GHz. Only 16 users could be provided with *Integrated Services Digital Network* (ISDN) services with a BER of 10^{-6}, giving rise to 8 ISDN-B channels. Extrapolating these results with respect to the 16 fold increased offered bandwidth of 56 MHz in the

Table 7.6 Literature based values for realistic CDMA channels

Service	v^2	$F(G)$
class 1	0.2	0.009
class 2	0.2	0.010
class 3	0.2	0.012

Table 7.7 Comparison of capacity in terms of number of connections for different service classes in different CDMA systems

Transmission Mode	Service	only I_{intra}	plus I_{inter}	plus multipath
	class 1	1082	1079	471
SC	class 2	347	344	227
	class 3	124	124	77
	class 1	1082	1079	471
MC	class 2	358	356	236
	class 3	140	139	90

FWA network, a number of 1024 32 kbit/s connections seem possible. This capacity is similar to the presented CDMA calculation results over good conditioned channels of 1079 simultaneous connections.

7.5   Comparison of TDMA and CDMA Results

A comparison of TDMA and CDMA technology is not directly possible. It must be stated here that the capacity analysis for TDMA is performed by means of simulations whereas the capacity analysis of CDMA is carried out on an analytical basis. Due to these two different approaches there are technology specific assumptions and simplifications which have to be made. However, a capacity comparison on a system level can be drawn. To make the comparison fair, both systems are given similar features. The details of the physical layer that means, transmission bandwidth, background noise, etc. are the same for both systems. The system scenario and the transmission power levels of the radio stations are also equal for both systems.

The FWA network employing TDMA with dynamic channel allocation is simulated with different capacity enhancing schemes. It is shown what steps can be taken to raise the amount of traffic that can be carried by one RBS, Chapter 8. The system configuration with power control and directional antennas at the RNT applied corresponds closest to the scenario for the analytical CDMA capacity calculation.

The capacity, e.g. the maximum number of connections that can be carried by the CDMA system is calculated using an analytical method. In this model, perfect power control was assumed. The maximum capacity is attained for a CDMA channel which is assumed to be ideal. An ideal channel in this case is characterized by the fact that no interference from other cells and no multipath propagation is present. Allowing interference will diminish the number of possible connections as the quality of the radio channel degrades. Thereby, the effect of inter-cell interference is negligible since directional antennas at the RNT sides are assumed and rather less interference is emitted to the other cell's receiving RBS. But the multipath component in radio propagation has a severe impact on the number of connections that can be established with a sufficient quality.

The capacity comparison is performed for connections with properties of the class 1 service which are voice calls with a service bit rate of 32 kbit/s.

For the ideal CDMA system a maximum of 1082 class 1 connections can be carried. This number is reduced to 1079 if inter-cell interference is considered.

A value for the number of class 1 connections in a corresponding TDMA system can also be calculated. Simulations are performed for an FWA network with a maximum data rate of 35 000 slots/s. The TDMA system with the level-based power control and the application of narrowbeam RNT antennas can carry 60 % of this maximum traffic. Now the following calculation can be done. Having set the length of a slot to 800 bit, a total of 16.8 Mbit/s can be carried by the multicellular TDMA system. Dividing this bit rate by the class 1 service bit rate of 32 kbit/s, the number of class 1 connections can be obtained as that 525 class 1 connections can be simultaneously carried. Comparing this value the capacity of CDMA would mean that TDMA can only offer half of the CDMA capacity.

However, this is not quite correct. In the analytical approach to estimate the CDMA capacity, a heavy influence of multipath propagation causing ISI and increased inter-cell interference has been seen. The number of class 1 connections per cell in a multicellular environment with realistic CDMA transmission channels is reduced to 471. Now the capacity of both systems is in an equivalent range.

Furthermore, in the TDMA scenario the entire bandwidth was divided into sub-bands and every RBS was only able to transmit on one frequency at a time. For the CDMA the whole frequency range was considered in the calculations.

Considering interference characteristics, the primarily comparison of TDMA with an ideal CDMA system is valid. The effects of multipath propagation which are neglected for ideal CDMA are not included in the simulative TDMA study either. But, since smaller frequency sub-bands and with that lower transmission bit rates are proposed, the effect of ISI is not expected to have such a significant impact as for CDMA transmission. Therefore, the ideal transmission channel should not be assumed in the analytical approach. A little surprising is the result that inter-cell interference plays a very small role in reducing capacity in a CDMA system. This can be explained, however, through the very high spreading gain that can be employed in the CDMA network and due to the high directional subscriber antennas.

The results of this comparison are conform to other studies that were carried out. In [8], for example, a comparative study between CDMA and TDMA has been done for a single class service. Here, the capacity of each system depends on the *Quality of Service* (QoS) stipulated for the service. CDMA performs better when the QoS metrics are strict because of its good tolerance to interference.

TDMA on the other hand is susceptible to interference and thus performs better for services with low QoS demands. In general, however, the number of users that are allowed into the system are the same for both technologies.

A study presented in (Wilson, N. D. and Ganesh, R. and Joseph, K. and Raychaudhuri, D., 1993) [18], shows that CDMA has a 2:1 capacity advantage over TDMA. The TDMA system here, however, is employed without dynamic channel allocation or any other capacity enhancing techniques. The study shows that CDMA performs better than TDMA for short data messages but worse for longer messages, which may be encountered during file transfer for example.

List of Abbreviations

BER	Bit Error Ratio
BRAN	Broadband Radio Access Network
C/I	Carrier to Interference ratio
CCI	Co-Channel Interference
CDMA	Code Division Multiple Access
ETSI	European Telecommunications Standardisation Institute
FWA	Fixed Wireless Access
GDP	Gross Domestic Product
GSM	Global System for Mobile Communications
IP	Internet Protocol
ISDN	Integrated Services Digital Network
ISI	Inter-Symbol Interference
ITU	International Telecommunications Union
LAN	Local Area Network
MADCAT	Mobile ATM Dynamic Channel Allocation simulaTor
MAI	Multiple Access Interference
MC	MultiCode
MPP	MultiPoint-to-Point
PDC	Personal Digital Cellular
PTO	Public Telephony Operator
QoS	Quality of Service
RBS	Radio Base Station
RNT	Radio Network Terminal
SC	Single Code
TDMA	Time Division Multiple Access
VoD	Video on Demand
WLL	Wireless Local Loop
WSSUS	Wide Sense Stationary Uncorrelated Scattering
xDSL	x (=generic) Digital Subscriber Line

References

[1] F. Adachi, M. Sawahashi and H. Suda, 'Wideband DS-CDMA for Next-Generation Mobile Communications Systems,' *IEEE Commun. Mag.*, pp. 56–69, Sep. 1998.

[2] J. S. Evans and D. Everitt, 'Effective Bandwidth-Based Admission Control for Multiservice CDMA Cellular Networks,' *IEEE Trans. Vehicular Technology*, vol. 48, pp. 36–46, 1999.

[3] E. Geraniotis and B. Ghaffari, 'Analysis of Direct-Sequence Spread-Spectrum Multiple-Access Communication over Rician Fading Channels,' *IEEE Trans. Commun.*, vol. 39, pp. 713–724, 1991.

[4] K. S. Gilhousen, I. M. Jacobs, R. Padovani, A. J. Viterbi, L. A. Weaver and C. E. Wheatley III, 'On the Capacity of a Cellular CDMA System,' *IEEE Trans. Vehicular Technology*, vol. 40, pp. 303–311, 1991.

[5] J. Haine, 'HIPERACCESS: an Access System for the Information Age,' *Electronics Commun. Engineering J.*, vol. 10, pp. 229–235, Oct. 1998.

[6] C. Hart, 'Fixed Wireless Access: a Market and System Overview,' *Electronics Commun. Engineering J.*, vol. 10, pp. 213–220, Oct. 1998.

[7] ITU Radiocommunication Bureau, *Wireless Access Local Loop*, Geneva, 1997.

[8] S. V. Krishnamurthy, A. S. Acampora and M. Zorzi, 'On the Capacity of TDMA and CDMA for Broadband Wireless Packet Access,' in *The Ninth IEEE International Symposium on Personal, Indoor and Mobile Radio Communications*, vol. 1, pp. 724–744, 1998.

[9] S. J. Lee, H. W. Lee and D. K. Sung, 'Capacities of Single-Code and Multicode DS-CDMA Systems Accommodating Multiclass Services,' *IEEE Trans. Vehicular Technology*, vol. 48, pp. 376–384, Mar. 1999.

[10] W. C. Y. Lee, 'Overview of Cellular CDMA,' *IEEE Personal Commun.*, pp. 49–54, Feb. 1998.

[11] A. May, 'Wireless Local Loop: Why the Slow Take Up?' *Electronics Commun. Engineering J.*, vol. 10, pp. 236–238, Oct. 1998.

[12] P. Mermelstein, A. Jalali and H. Leib, H. 'Integrated Services on Wireless Multiple Access Networks,' in *IEEE Proceedings Vehicular Technology Conference*, vol. 2, pp. 863–867, 1996.

[13] T. S. Rappaport, *Wireless Communications: Principles and Practice*, Prentice Hall PTR, Upper Saddle River, NJ, 1996.

[14] J. E. Smee and H. C. Huang, 'Mitigating Interference in Wireless Local Loop DS-CDMA Systems,' in *The Ninth IEEE International Symposium on Personal, Indoor and Mobile Radio Communications*, vol. 1, pp. 724–744, Sep. 1998.

[15] B. Steiner and P. Kuhlmann, 'Downlink Performance of a DS-CDMA Wireless Local Loop System,' in *IEEE Proceedings Vehicular Technology Conference*, vol. 4, pp. 1331–1335, 1999.

[16] B. Walke, *Mobile Radio Networks*, John Wiley & Sons, Chichester, UK, 1999.

[17] W. Webb, 'A Comparison of WLL with Competing Access Technologies,' *Electronics Commun. Engineering J.*, vol. 10, pp. 205–212, Oct. 1998.

[18] N. D. Wilson, R. Ganesh, K. Joseph and D. Raychaudhuri, 'Packet CDMA Versus Dynamic TDMA for Multiple Access in an Integrated Voice/Data PCN,' *IEEE J. Selected Areas Commun.*, vol. 11, pp. 870–884, 1993.

[19] World Bank, *Knowledge for Development*, Oxford University Press, New York, 1999.

8

Traffic based Dynamic Channel Allocation Schemes for WLL

Ingo Forkel, Stefan Mangold, Roger Easo and Bernhard Walke

8.1 Introduction

The need for providing telecommunication services is growing faster than ever. For this purpose *Wireless Local Loop* (WLL) also known as *Fixed Wireless Access* (FWA) Networks is an effective alternative to the problematic wired system. Compared to mobile systems, FWA networks provide two-way communication services to near-stationary users within a small area [8].

The next section presents an overview of multiple access schemes. Channel allocation strategies are described in the second part of the section. The emphasis here is on fixed and dynamic channel allocation schemes. A unique *Dynamic Channel Allocation* (DCA) technique for a *Wireless Asynchronous Transfer Mode* (WATM) system based on the *Carrier to Interference ratio* (C/I) measuring is presented. This technique is explained in detail as it provides the concept for the simulator which is used for the analysis of the broadband FWA network.

The FWA network scenario and its parameters are presented in the following section. The parameters of the radio channel are explained as well as the traffic models and the system parameters such as its geometrical setting and radio characteristics.

Simulation results of *Time Division Multiple Access* (TDMA) with DCA are presented. The effects of different capacity enhancing schemes suitable in WLL scenarios for the proposed algorithm are investigated and evaluated.

8.2 Access and Allocation Technologies

Presented in this section is a description of multiple access technologies and an overview of existing channel allocation schemes. An extended version with some examples explained in more detail is given in [5].

8.2.1 Multiple Access Technology

A defined radio bandwidth can be divided into a set of defined radio channels. Each channel can be used simultaneously while maintaining an acceptable received radio signal.

The radio spectrum can be divided into separate channels using splitting techniques such as *Frequency Division* (FD), *Time Division* (TD) or *Code Division* (CD) multiple access. Let $S_i(k)$ be the set i of wireless terminals that communicate with each other using the same channel k. By taking advantage of the radio propagation loss, the same k channels can be reused by another set j if both, i and j are spaced sufficiently apart. All sets which use the same channel are known as co-channels. The co-channel re-use distance σ denotes the minimum distance for re-use of the channel with an acceptable level of interference. A channel can be reused by a number of co-channels if the C/I in each co-channel is above a required minimum C/I. C represents the received signal power in a channel and I is the sum of all received signal powers of all the co-channels.

Consider the scenario depicted in Figure 8.1.

Here, an *Radio Network Terminal* $(RNT)_t$ is transmitting to its *Radio Base Station* (RBS) located at a distance d_t. Five surrounding RNTs, communicating with their respective RBS located at distances $d_1 \ldots d_5$ on the same channel as RNT_t, cause interference at RBS_t.

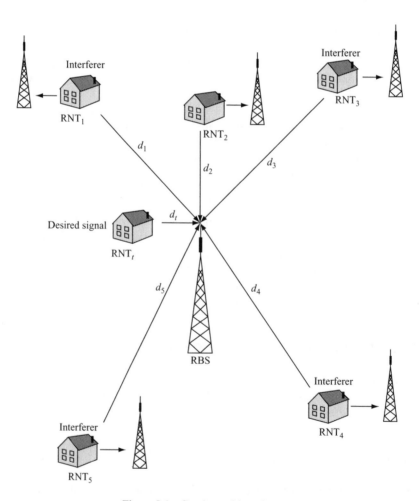

Figure 8.1   Co-channel interference

Denoting the transmission power of the RNTs with $P_{i=t,1,...,5}$ then Equation (8.1) describes the co-channel interference caused at the reference RBS in an abstract form. N here is the background noise and α is the propagation coefficient which is determined by the terrain

$$C/I = \frac{P_t d_t^{-\alpha}}{\sum_{i=1}^{5} P_i d_i^{-\alpha} + N} \tag{8.1}$$

As Equation (8.1) shows there are different methods of obtaining a satisfactory C/I at the reference station RBS. For example, the distance between the co-channel stations can be increased, or the interfering power can be reduced (or increasing of desired signal power P_t). The latter describes the motivation underlying the power control schemes.

8.2.2 Channel Allocation Technology

Channel allocation schemes can be divided into *Fixed Channel Allocation* (FCA), *Dynamic Channel Allocation* (DCA) and *Hybrid Channel Allocation* (HCA). These allocation schemes are based on the method in which the co-channels are separated.

8.2.2.1 Fixed Channel Allocation

In an FCA strategy neighbouring radio cells are grouped together to form clusters. The total number of available channels is divided into sets. Each radio cell in a cluster is allocated a set of these radio channels. The number of channels and with that the cluster size within the set is dependent on the frequency re-use distance and the required signal quality.

Considering a hexagonal cell with radius R and a distance D between the cluster centres, the minimum number n of channel sets necessary to cover the FWA network area is

$$n = \frac{1}{3}\sigma^2 \quad \text{whereby} \quad \sigma = \frac{D}{R} \tag{8.2}$$

For $\sigma = 3$, the minimum number of radio channel sets is $n = 3$. In the simple FCA strategy the same number of radio channels is allocated to each cell. This distribution is efficient for a uniform traffic load distribution in the FWA network system. In this case the blocking probability in one cell is the same as the blocking probability of the whole system. FCA strategies require careful planning of the distribution of channels and cells.

The implementation of FCA is simple since it is a static system. However, FCA reaches its limit when it has to serve a varying traffic load in the FWA network system.

FCA encounters a problem. If the traffic distribution within the radio system is uneven, it can happen that the blocking probability in heavily traffic loaded radio cells quickly reaches a maximum despite free channels which are present in lightly loaded radio cells. The resulting effect is a poor channel utilization. To improve this utilization either non-uniform channel allocation or channel borrowing schemes may be applied.

8.2.2.2   Dynamic Channel Allocation

As opposed to FCA, in DCA there is no exclusive association between channel and the radio cell. As a result, DCA strategies are able to react flexibly to local and temporal variations of mixed traffic and load distributions [5]. A radio channel is used by any radio cell as long as signal interference constraints are met. There are three major types of DCA strategies, centralized, decentralized, and C/I measurement based schemes.

In Table 8.1 an overview of the different strategies is presented. For FWA networks the schemes *Dynamic Channel Selection* (DCS), already in use within the *Digital Enhanced Cordless Telephone* (DECT) system, and *Channel Segregation* (CS) are believed to have the greatest importance. However, the schemes must be adapted for the FWA scenario.

8.2.3   Dynamic Channel Allocation for FWA Networks

The FWA network scenario must support a packet-oriented environment like ATM. Since ATM implies different *Quality of Service* (QoS) and bandwidth requirements, the implementation of DCA as a channel assignment strategy seems appropriate. As mentioned before DCA is able to react flexibly to fluctuations in traffic. For real-time services classes like *Variable Bit Rate* (VBR) there must always be enough capacity to guarantee transmission of the cells.

All channel allocation strategies are based on the assignment to physical channels. The various channel access protocols perform the characteristic statistical multiplexing of ATM cells of RNTs in the service area of an RBS. The channel access is co-ordinated by the RBS. The virtual connections of the RBS to the RNTs occupy the entire frequency spectrum.

The problem of the capacity allocation is to find a method to give each RBS sufficient capacity. An RBS should only be given that portion of the frequency which is really necessary, while the remaining part is delivered to the other RBSs. However, the allocation dynamic must be held minimal to avoid interference to the other RBSs.

Table 8.1   Overview of DCA schemes

Category	Scheme
Central DCA	First Available
	Locally Optimized Dynamic Assignment
	Selection with Maximum Reuse Ring Usage
	Mean Square
	Nearest Neightbour
	1-Clique
Distributed DCA	Locally packing distributed DCA
	Moving Direction
C/I measurement based DCA	Sequential channel search
	Minimum Signal-to-Noise Interference Ratio
	Dynamic channel selection
	Channel segregation

The studied system has a carrier frequency of 28.5 GHz with a bit rate of up to 112 Mbit/s if *Quaternary Phase Shift Keying* (QPSK) modulation is considered. The available spectrum can be divided into channels, corresponding to the number of frequencies set for the spectrum. The cell radius of the FWA network lies in the range of up to 2500 m. The allocation of a complete frequency as a physical channel is uneconomical as the RBS does not require the complete resource all the time. Hence, the idea of splitting the total capacity within a frequency band into multiple resource units by time division multiplex seems appropriate.

DCA Principle Following is a method to implement a DCA scheme for WATM in for FWA. The problem which is characteristic of DCA is that the scheme requires a steady behaviour whereas ATM load is synonymous for dynamic behaviour. The technique presented here is based on a C/I measuring DCA scheme.

The Medium Access Protocol The *Medium Access Control* (MAC) protocol applied here is very much like the European HIPERLAN/2 system and a candidate for the HIPER-ACCESS standard. It controls the access of the RNTs and the RBSs to the shared radio channel. The access is co-ordinated by the RBS which acts as a central instance. The RBS allocates capacity to the RNTs on a slot by slot basis. The scheduler within the RBS does not have any direct information on the waiting buffers of the RNTs. Rather, the RBS contains a mirror of each RNTs occupancy state of the send buffer. This information is sent from the RNT to the RBS via a signalling scheme. The MAC protocol divides the transmission into signalling periods. The length of the signalling period can be variable, however, here it is of a fixed length. The signalling period of the transmission can be divided into four phases as depicted in Figure 8.2.

The four different transmission phases in each fixed length signalling period are explained in the following:

- *Broadcast phase*: This phase consists of the broadcast control channel which contains general information and is sent at the beginning of each frame. The phase also consists of a frame control channel which contains information on the next frame.
- *Downlink phase*: During the downlink phase control units and data units are sent from the RBS to the respective RNTs.

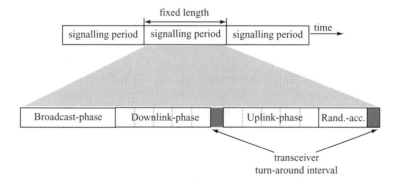

Figure 8.2 Structure of the MAC frame

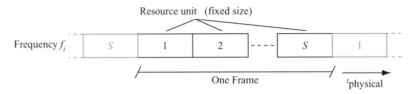

Figure 8.3   Frame structure of the physical channel

- *Uplink phase*: The uplink phase is similar to the downlink phase with the exception that the transmission is from the RNTs to the RBS.
- *Random access phase*: The RNTs send control information to the RBS in a contention based manner, which means that collisions can take place.

Method   A frequency is divided into intervals of equal length using TDMA. A fixed number of intervals S (now called resource units) are grouped together to form a frame. In the course of the simulations the size of a resource unit was set to 5 slots (5 ATM cells), giving a duration of 100 µs.

The channel allocation for an RBS takes place during a resource unit (or several resource units) according to the capacity required by the RNTs' connections. A frame is repeated periodically, thereby creating a steady behaviour, since an RBS allocates the same resource unit in every frame. In effect, a new physical channel has been created. The allocation takes place on the basis of interference measurements. The measurements are done on many frames but always on the same resource unit for each RBS.

The MAC protocol is in charge of co-ordinating the access of the RBS and the RNTs to the channel within the allocated resource unit(s). As a result, within the resource units a dynamic capacity allocation takes place according to the need expressed by the RNT.

8.3   FWA Network Scenarios

This chapter contains a description of the FWA network scenario and its parameters. This is also the basis for the capacity analysis between simulative TDMA and analytical *Code Division Multiple Access* (CDMA) in Chapter [Ref Chapter CDMA vs TDMA].

8.3.1   Basic Parameters of the FWA Network Scenario

System Setup   The FWA network simulations are carried out with a hexagonal cell scenario containing 61 cells. One RBS is placed in the centre of each cell and 48 RNTs are uniformly distributed over each single cell area. The inner 19 cells are evaluated and the outer two rings of cells produce additional interference to achieve a realistic traffic and frequency usage for the simulation.

Cell Sizes   By means of experiments conducted in San Francisco and nearby areas it was determined that the optimal FWA network could have a cell size radius of approximately

2 km [12]. The simulations will be carried out using cells with this radius. Table 8.2 summarizes the basic system parameters.

8.3.2 Radio Aspects

Spectral Efficiency and Available Bit Rate   According to German spectrum regulations [11], the available bandwidth per FWA-operator at 26 GHz will be 56 MHz. Assume the values in Table 8.3, in order to define the spectral efficiency of the particular modulation schemes with variable bit rates.

Using this basic spectrum efficiency model, the available bit rate, that is supported by the system is calculated to 112 Mbit/s for QPSK modulation, and if 16 *Quadrature Amplitude Modulation* (QAM) is used 224 Mbit/s [10].

If the frequency band is divided into sub-carriers, the available bandwidth in the sub-bands leads to a certain transmission speed. The TDMA simulations are carried out for four sub-bands within the entire bandwidth available for the DCA and FCA evaluations. Assuming the length of a MAC-*Packet Data Unit* (PDU) on the physical layer to be about 800 bit (including coding, and protocol overhead) a total of

$$\frac{112 \text{ Mbit/s}}{800 \text{ bit/cell} \times 4 \text{ frequencies}} = 35\,000 \text{ cells/s/frequency} \qquad (8.3)$$

can be carried using QPSK modulation. The duration of a slot carrying one MAC-PDU with one ATM cell as payload is then 28.6 μs. The use of a higher-order modulation

Table 8.2 Simulation parameters

Parameter	Value
Cells	61 (inner 19 evaluated)
Cell radius [km]	2.0
RBSs per cell	1
Sectors per cell	1[1] or 6[2]
RNTs per sector	48[1] or 8[2]

[1] Omni-directional RBS antenna
[2] Sectorization at the RBS applied.

Table 8.3 Spectral Efficiencies

Modulation	Spectral Efficiency[1]	Available Bit Rate	Slot Duration
QPSK	2.0 (1.5) b/s/Hz	112 (84) Mbit/s	28.6 μs
16 QAM	4.0 (3.5) b/s/Hz	224 (196) Mbit/s	14.3 μs

[1] Basic example values proposed in (NORTEL NETWORKS, 1999) in brackets.

scheme results in higher transmission bit rates corresponding to smaller slot duration. Also in the case the modulation order is to be changed, the resource unit duration must be kept constant. The number of slots per resource unit would then be increased.

Propagation Model   In the propagation model employed in the simulator a transmission coefficient n models the propagation in different environments like urban, sub-urban, residential, or hilly terrain. Based on measurements in German urban areas $n = 2.7$ has been selected [1].

The path loss model implemented in the simulator is described by

$$L_{\text{PL}}(d) = L_F(d_0) + 10n \log(d/d_0) \tag{8.4}$$

where n is the propagation coefficient discussed before. The reference path loss value $L_F(d_0)$ in a distance $d_0 = 1 \, \text{m}$ is calculated based on the free space path loss formula

$$L_F(d) = 20 \log \left[\frac{4\pi d}{\lambda} \right] - G_{\text{Antennas}} \tag{8.5}$$

with wavelength and the gains of the transmitter and receiver antennas. Hence, the reference path loss is $L_0 = 61.55 \, \text{dB}$. This leads to a maximum path loss in distance $d = 2 \, \text{km}$ at the edge of coverage of $L_{\text{PL,max}} = L_{\text{PL}}(d = 2 \, \text{km}) = 150.67 \, \text{dB}$. The path loss of all stations to each other are based on this path loss model.

Additionally, a fading margin of 3 dB is included in the propagation model to acknowledge the fading effects caused by multipath propagation and signal shadowing.

Background Noise   Considering a background noise power spectral density of $N_0 = 4 \, \text{pW/GHz}$ a noise level of $-97.5 \, \text{dBm}$ for the frequency band of 56 MHz is found. If the whole bandwidth is divided into four sub-bands, each frequency band is disturbed by a noise level of $-102.5 \, \text{dBm}$.

Additional noise is caused by the receiver components. A receiver noise figure of 5 dB is considered in the simulations, taking into account the high quality devices applicable at the FWA network subscriber radio units and the RBSs. Including the receiver noise figure a background noise level of $-97.5 \, \text{dBm}$ is finally adopted.

Root Mean Square Delay Spread   Measurements carried out at the University of Paderborn for an FWA scenario at 29.5 GHz revealed that multipath at this carrier frequency practically does not exist [4]. This may be due to the very strong *Loss of Sight* (LoS) component between the RNT and RBS, or due to technical limitations in the measuring equipment. In fact an RMS delay spread of DS $= 1 \, \text{ns}$ can hardly be measured.

Transmission Power   The transmission powers of the RBS and the RNT depend on eventual antenna gains, the attenuation of the radio link between RBS and RNT and the background noise level. Considering the background noise level at $-97.5 \, \text{dBm}$, the maximum path loss of $L_{\text{PL,max}} = 150.67 \, \text{dB}$, the assumed fading margin of 3 dB, and an appropriate SNR of 15 dB, the necessary maximum transmission power of an RNT or RBS will be in the range of up to 71 dBm if there are no antenna gains (see Table 8.5). It can be reduced by both, the transmitting and receiving antenna gain value since the path loss value will decrease by these gain values.

Power Control For FWA networks, it will be suitable to apply power control to the RBS and RNT transmitters, at least for the uplink point-to-point data transmission. The transmission power calculation presented in the last section refers to full cell site coverage. RNTs at locations close to the RBS can work with significantly lower signal powers and reducing their transmitter power and the respective power at the RBS will reduce the overall system interference.

The simulations will consider power control depending on the path loss value of the link between the RNTs and their belonging RBS. Hence, the transmitting power $P_{Tx(PC)}$ with power control applied is individually reduced to

$$P_{Tx(PC)} = \min\{P_{\max},\ P_{\text{req}} + L_{PL}(d = d_{\text{Link}})\}, \tag{8.6}$$

where $P_{\max}$ is the maximum transmitting power to overcome the maximum path loss at the cell edge and $L_{PL}(d = d_{\text{Link}})$ is the path loss value of the respective link. As a result all the received signals have the same power level at the receiver as long as the maximum transmitting power offers the required level [3].

Received Signal Power The FWA network being analysed for direct sequence CDMA is assumed to behave power controlled. This means that in every radio cell each RNT signal arrives at its respective RBS with the same received power level. The received signal power value of $-78\,\text{dBm}$ is assumed for the three different scenarios to be investigated. This value is taken over from the equivalent TDMA FWA network simulation campaign for the same scenario.

Modulation Following the proposal of the *Digital Audio Video Council* (DAVIC) for the *Local Multipoint Distribution System* (LMDS) transmission technology, two modulation schemes shall be applied in FWA networks, QPSK and 16 QAM. The choice of the modulation technique to be employed will depend on the current link quality, e.g. the *Signal-to-Noise Ratio* (SNR) of the connection. For sake of simplicity the modulation technique here is restricted to QPSK.

Interference The system itself or systems of other operators using the neighbouring frequency range causes interference which can be treated as additional noise.

Interference consists of *Inter-Channel Interference* (ICI), *Adjacent Channel Interference* (ACI), and *Co-Channel Interference* (CCI). The first is a result of delayed signal parts which are caused by multipath propagation or failed synchronization of the receiver. ACI considers the overlapping parts of the frequency spectrums of neighbouring frequency channels, either in the same system or of different systems. CCI is calculated during the simulations considering all current connections established within the same time interval and frequency channel. Therefore, a carrier signal power has to be provided significantly higher than theoretically derived from the link level, e.g. coding performance analysis to offer a sufficient *Packet Error Ratio* (PER) over the whole radio cell.

Inter-Cell Interference The calculation of the interference caused by neighbouring radio cells to a reference cell is necessary for the capacity equations developed for the multi-cellular CDMA system.

The inter-cell interference can be computed using the geometric position of the RNT and RBS in the scenario when power control is applied. The summation of these values give the total other cell interference present in the reference cell. This I_{inter}, however, results from the assumption that all terminals are always active. This is not quite correct as the terminals defined by the traffic model presented below are only active for a certain time T_{on} and they are idle during the time T_{off} otherwise. During T_{on}, each RNT generates cells at a mean cell rate of 5%. This activity of the RNT determines the interference it causes.

$$I_{inter} = \sum I_{ij} \cdot 5\% \cdot \frac{T_{on}}{T_{off} + T_{on}} \qquad (8.7)$$

The I_{inter} values in Table 8.4 are calculated for scenarios with different antenna arrangements. The I_{inter} values are relative to the received signal power P_i of a power controlled RNT to its serving RBS in the reference cell. The values for the time intervals deciding the activity are taken for the equivalent TDMA FWA network with 30% average traffic offer per RBS in a radio cell.

Antenna Patterns   The FWA network fixed-to-fixed radio link allows the deployment of high-gain directional antenna at the user's houses as well as sectored antennas at the base stations. This can improve signal strength and reduce interference. The following antenna characteristics summarized in Table 8.5 will be used in the investigation.

Due to the use of directional or beam antennas at the RNT side, only a few RNTs of a given scenario cause interference referring to a specified location, either an RBS or RNT. Suppose the additional loss $L_{Antenna}(\alpha)$ of an RNT antenna with beamwidth $\delta = 2\delta_0$ is given by

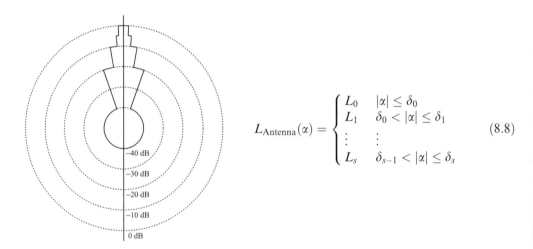

$$L_{Antenna}(\alpha) = \begin{cases} L_0 & |\alpha| \leq \delta_0 \\ L_1 & \delta_0 < |\alpha| \leq \delta_1 \\ \vdots & \vdots \\ L_s & \delta_{s-1} < |\alpha| \leq \delta_s \end{cases} \qquad (8.8)$$

where is the angle between the *Line of Sight* (LoS) path, which will be the antenna beam justification, and the direct link line to another terminal. $L_0 = 0\,$dB should be convenient, since the antenna gains G_{RNT} or G_{RBS} of the link participating stations are already included in the free space propagation formula Equation (8.5), and the s values for L_i and δ_i, respectively, should consider the beam form of the RNT antenna as depicted in Equation (8.8). Note that for increasing s the modelling of the RNT beam antenna becomes more detailed as the value of s defines the number of degree intervals within the antenna pattern associated with different losses L_i with $i = 1, \ldots, s$.

For exact simulations the number and position of all active terminals at a given moment have to be taken into account to calculate the currently engendered interference. Hence, the propagation path loss L_{PL} of each link between two terminals is amended by the beam antenna characteristic of the sending as well as the receiving terminal, considering the angles α_1 and α_2 between that link and both connections to the terminals associated RBS(s).

Table 8.4 Total other cell interference

Scenario	T_{on}	T_{off}	I_{ij}	I_{inter}
Omnidirectional	10 s	70 s	505.62	3.096 P_i
Narrowbeam	10 s	70 s	13.81	0.086 P_i
Sectored	10 s	70 s	11.60	0.073 P_i

Table 8.5 Antenna characteristics

	RBS Antenna		
	Omnidir.		Sectored
Parameter	RNT Antenna		RNT Antenna
	Omnidir.	Direct.	Direct.
RBS Tx Power [dBm]	71	42	30
RBS Antenna Gain [dBi]	0	0	12
δ_0, L_0	–	–	30, 0
δ_1, L_1	–	–	35, 5
δ_2, L_2	–	–	40, 10
δ_3, L_3	–	–	50, 20
δ_4, L_4	–	–	180, 40
RBS Tx Power [dBm]	71	42	30
RBS Antenna Gain [dBi]	0	29	29
δ_0, L_0	–	3, 0	3, 0
δ_1, L_1	–	5, 5	5, 5
δ_2, L_2	–	10, 10	10, 10
δ_3, L_3	–	20, 20	20, 20
δ_4, L_4	–	180, 40	180, 40

Table 8.6 summarizes the system parameters considering transmission techniques and radio aspects.

8.3.3 Coding Aspects

Following the proposal of DAVIC for LMDS systems [2], ATM data cells should be protected with a *Forward Error Correction* (FEC) mechanism. A shortened *Reed-Solomon Code* (RSC) is applied to each packet data unit. For further information please refer to S. Mangold and I. Forkel [7].

8.3.4 Traffic Aspects

The comparison of TDMA and CDMA is based on two different traffic models. The simulated TDMA system is a packet switched system whereas the analysis for CDMA is centred around a circuit switched system.

8.3.4.1 TDMA Packet Switched Traffic Model

In order to enable the modelling of realistic ATM traffic within the simulated multi-cellular FWA networks, the characteristics of the used PDU-sources are discussed in this section.

Each RNT is defined by a fixed position, its link to one RBS and various statistical traffic parameters. In order to understand the outcome of the simulations and the

Table 8.6 Radio parameters

Parameter	Value
Frequency [GHz]	28.5
Bandwidth [MHz]	56
Frequency Bands	4
Modulation	QSPK
Propagation Coefficient	2.7
Path Loss at 1 m [dB]	61.55
Path Loss at Edge of Coverage [dB]	150.67
Fading Margin [dB]	3.0
Receiver Noise Figure [dB]	5.0
Background Noise [dBm]	−97.5
Adjacent Channel Interference [dB]	$-60 \cdots -40$
Received Signal Power [dBm]	−78

manifold effects of the resource allocation, cell scheduling, multiplexing, cell delays, congestion between the competing RNTs at RACH and various other techniques, *Available Bit Rate* (ABR) traffic only will be modelled rather than mixed multimedia services with *Constant Bit Rate* (CBR), VBR, ABR and *Unspecified Bit Rate* (UBR) service classes. ATM cells and their respective MAC-PDUs are generated for uplink and downlink, where at each single RNT the traffic is modelled as a full uplink or downlink stream, not mixed. The asymmetric offer will be provided by allowing both types of RNTs communicating simultaneously. The different activities of the individual RNTs define the asymmetric offer per RBS.

An asymmetric average offer of 1:4 (up/down) as well as a symmetric average offer of 1:1 will be simulated. A cell arrival means also the arrival of one MAC-PDU at the RNT or RBS queues and is referred to as *Cell Reference Event* (CRE).

Figure 8.4 illustrates a scenario configuration with four simultaneously active RNTs, two with uplink and two with downlink MAC-PDUs.

An RNT is characterized by one ABR connection, alternating between active and idle period times. A generative state model is applied to define these random period times, see Figure 8.5. Both, active and idle times are negative exponential distributed random numbers. Typical mean times are $T_{on} = 10$ s and a varying T_{off} between 1 s and 150 s, where T_{on} is the time the model is in state active and T_{off} means the time the model is in state idle. Applying traffic models for the RNTs in which terminals are sometimes active

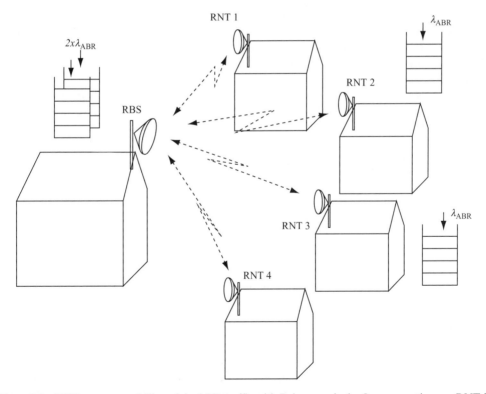

Figure 8.4 PDU sources modelling plain ABR traffic with Poisson arrivals. One connection per RNT is assumed with either uplink or downlink traffic

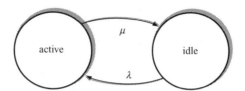

Figure 8.5   Traffic model for one RNT

and sometimes silent, leads to time-varying interference situations and capacity demands where hot spot areas occur randomly on arbitrary locations.

During active phases, the connection is characterised by Poisson arrivals of cells in up-or downlink. All stochastic generators for these CREs work independently of each other.

The traffic offer must be varied within the simulations in order to find and determine the system capacity limits with FCA and DCA schemes. In the ˙simulations, this will be ensured by varying the mean idle time of all RNTs. Each connection is defined to use a certain amount of the maximum available capacity of one RBS which is defined as the maximum capacity of one frequency. Note that at the RBS one single transceiver is used which means that only one frequency channel can be operated simultaneously.

The overall system capacity highly depends on the fact that a connection is dropped if a certain threshold in the PER, e.g. 5 % is exceeded. As a matter of fact, the gain from statistical multiplexing is reduced if a small number of active connections (or active RNTs) per RBS, each with a relatively high traffic offer, is assumed rather than a large number of connections where the individual connections contribute to the overall traffic only a little [6].

The following simulation parameters of Table 8.7 define the traffic sources.

If the downlink to uplink ratio is 1:1, all RNTs are assigned to the same mean idle time T_{off}. The offer per RNT, i.e. the offer per virtual connection with respect to the total offer to one radio cell is given by the following equation. The equation is still valid when sectored cells are introduced.

$$\lambda_{ABP}(RNT) = \frac{T_{on}}{T_{on} + T_{off}} \cdot MCR \qquad (8.9)$$

8.3.5   Parameters of the Resource Unit DCA Technique

The channel allocation technique of the simulations is based on the resource unit principle that was already presented above The parameters of the resource unit principle are found in Table 8.8. These parameters will be constant for all simulations. However, the duration of a slot can be modified to acknowledge the effect of the modulation schemes QPSK or 16 QAM.

The tolerable packet error ratio is set to 5 %. If this value is surpassed, the RBS tries to allocate a new resource unit for a better transmission. If a resource unit replacement cannot be found the *Cell Loss Ratio* (CLR) increases and finally connections are dropped. To avoid too many erroneous transmissions an *Automatic Repeat reQuest* (ARQ)

Table 8.7 Parameters for the traffic sources

	Source	Remark
Type	ABR	applies ARQ, typical multimedia
max. CTD	30 ms	minor real time demands
Generator	Poisson	uncorrelated arrivals
	State Model	Remark
mean T_{on}	10 s	Negative exponential distributed
mean T_{off}	1 s ... 150 s	random times defines traffic offer of particular connection
	ATM Cell Reference Events	Remark
Cell rate per conn.	up to 1633/s	48 RNT/radio cell lead to overload if all active
MCR[1]	5 % (during active periods)	in percentage of overall system load
MCR$_{min}$	0.9 MCR	the individual connecttion vary
MCR	1.1 MCR	slightly in their MCR characteristics

[1] Mean Cell Rate.

Table 8.8 DCA parameters

Parameter	Value
Frequencies	4
No of Resource Units per Frame	21
Resource Unit Length [slot]	5
Slot Duration [μs]	28.6
Transceiver Turn Around Time	0
FCA Cluster Size	7 or 12
MAC signalling period length [slot]	15
Interference Threshold [dBm]	−87

protocol can be switched on at the beginning of a simulation campaign. The ARQ protocol can improve the CLR at the cost of increasing the cell delay.

The reservation and allocation of a resource unit is performed by the RBS. The RBS measures the interference level of the resource unit it wants to employ. If the measured interference level is above a threshold value, the RBS considers the resource unit to be already in use and a replacement resource unit is searched for. The interference threshold

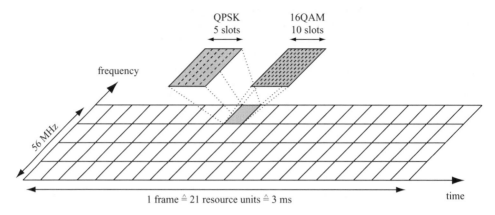

Figure 8.6   Resource unit principle of the DCA scheme

value selected for the simulations is determined by the transmission powers, the attenuation due to the path and the background noise.

The structure of the physical channel or, in other words, the TDMA frame configuration is presented in Figure 8.6. Also depicted is the influence of the modulation scheme. Based on this model for the physical channel, the simulations of the DCA scheme for FWA networks have been performed.

8.4   Simulation Results of DCA in FWA Networks

Presented here are the results of the simulations performed for DCA in FWA networks with the parameters defined above. After a brief description of the evaluation criteria, different enhancing technologies to improve the DCA performance are introduced and their effect on the system is commented. At the end of the chapter a capacity analysis is performed based on the quality of service offered by each system.

8.4.1   Criteria for Evaluation of the Simulation Results

To determine the quality of DCA in FWA networks, measurements of the system behaviour and the *Grade of Service* (GoS) with respect to blocked and dropped MAC connections are assessed. In addition, the cell loss ratio, cell transfer delays and when appropriate the C/I ratios are evaluated.

At each terminal, individual carrier and interference values are measured leading to a particular loss ratio of MAC-PDUs. One lost MAC-PDU results in the loss of a single ATM cell and assuming that retransmissions are required if ARQ is used, this will increase the transfer delay of the respective cell. The loss ratio depends on the error characteristics and the applied transmission and coding techniques, in other words, on the characteristics of the physical layer.

8.4.2   Simulation Results

During the initial simulations, the simulation parameters had to be verified. Of particular interest were the optimum transmission powers of the stations and the interference threshold upon which the availability of a free resource is determined.

Conducting a simulation strategy based on analytical assumptions without considering any antenna gains at the station sites the following values were attained:

- Transmission power of 71 dBm at both, the RNT and RBS.
- Interference threshold value of −87 dBm.

This transmission power is unrealistic as GaAs amplifiers cannot dispend such a high energy. But the application of realistic antennas allow to operate the transceivers at much lower power levels than assumed in this primarily simulations.

Following now are the results of a simulation campaign in which an attempt was made to analyse different capacity enhancing technologies for radio transmission in FWA networks.

8.4.2.1   FCA Simulation

FCA simulations were carried out with cluster sizes of 7 and 12. Figure 8.7 shows the scenario with a 7 cluster. The darker shaded region shows the radio cells that share the same radio resource units. Since there are 4 frequencies each with 21 resource units, 12 resource units are assigned to each radio cell among the seven radio cells in the cluster. Since 21 resource units are available in each frequency band accessible with one transceiver, the maximum achievable traffic load in the system is around 55%. Such a high load cannot be reached with the chosen dynamic traffic models and is also reduced by system interference.

Furthermore, a cluster size of 12 as illustrated in Figure 8.8 was selected for the simulations based on the following premise. Assuming that an RBS is able to carry a

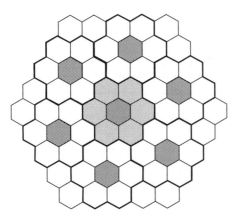

Figure 8.7   Scenario with cluster size 7

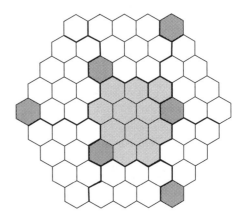

Figure 8.8   Scenario with cluster size 12

traffic load of 33 %, an RBS would require a capacity of almost one third of a frequency band, hence 3 RBSs could share one frequency band. Since the system is divided into 4 frequency bands, 12 RBSs could be served simultaneously accessing the overall system capacity which results in a 12 cluster. Dividing the resource among the radio cells of a 12 cluster, each radio cell is assigned 7 resource units. Indeed, the simulation campaign with a cluster size of 12 showed slightly better results than the counterpart with a 7 cluster.

Notable for these simulations was the very low traffic load that could be carried by the systems. Due to the high dynamic of the applied traffic models, the FCA prematurely reaches its capacity limit since all available resources have already been allocated. That happened also at a comparatively low amount of offered traffic. An increased number of blocked connections results out of this. Moreover, since FCA has no mechanism to provide transmission quality to the connections, a high PER has been noticed in the simulations for the smaller cluster size 7. Because of the shorter re-use distance, the co-channel interference increases in this scenario compared to the cluster size 12 system.

8.4.2.2 Simple DCA Scheme

These primary DCA simulations were carried out using the basic DCA scheme parameters explained in the section before. Compared to the FCA simulations there was a considerable increase in traffic that could be carried in the system.

With a traffic load of 26 % per RBS all connections were admitted into the system and all data packets were carried without transmission errors. Increasing the traffic load leads to a degradation of the system quality as proven later. Consider the number of rejected and aborted calls given in Table 8.9.

The effect that there are more aborted calls than rejected calls can be explained by the nature of the DCA technique. A connection is rejected only if the interference level measured on the resource unit is above the threshold value defined in the DCA parameter table. This threshold is a free parameter and is here set up to maximize the number of connections, on the expense of the larger number of dropped resource units. If a connection is established it can happen that the quality of the resource unit is so poor that the packet error rate exceeds the tolerance level of 5 %. The allocating entity (here the RBS) searches for a replacement for this resource unit, dropping the connection if the search fails.

Table 8.9 Simulation results for the simple DCA

Offered Traffic	Blocked Calls	Aborted Calls
26 %	0.00 %	0.00 %
28 %	0.00 %	0.02 %
30 %	0.00 %	0.06 %
32 %	0.02 %	0.07 %
34 %	0.01 %	2.04 %

The nature of the properties of the DCA scheme with varying traffic load are shown in the figures below. As shown in Figure 8.9 the SNR or C/I level decreases with increasing traffic. This is evident since the transmission power of the RNTs is constant whereas the interference is raised through the increased activity of the transmitting stations. Similarly the number of allocated containers increases if more traffic has to be carried. This effect is seen in Figure 8.10.

8.4.2.3 DCA with ARQ

One method to improve the quality of the DCA scheme with respect to CLR is the application of an ARQ protocol. The ARQ protocol requests the retransmission of a

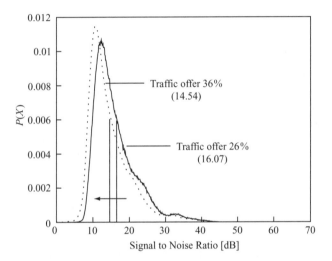

Figure 8.9 SNR for DCA uplink

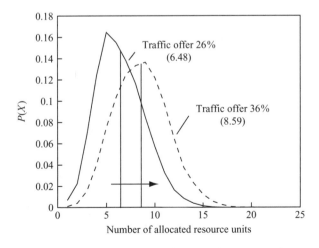

Figure 8.10 Mean number of allocated resource units

packet if it is identified as being faulty at the receiver. Thus, ARQ improves the CLR at the site of the receiver. However, since packets must be re-transmitted, the waiting time for packets to be serviced increases. This can lead to longer cell delays, a situation which might be intolerable for time critical service classes like CBR which have stringent values for the maximum tolerable cell delay. Comparing the cell delays of the previous simulation and the one with ARQ there is a notable difference. Figure 8.11 shows the juxtaposition of the cell delays for a traffic load of 34 % at the RBS. The simple DCA scheme has a mean cell delay of 51 slots which equals 1.46 ms. Moreover, the probability for delay values larger than 20 ms can be neglected. The DCA scheme with an ARQ protocol has a mean cell delay of 120 slots corresponding to 2.92 ms where almost 3 % of the packets exceed a delay of 20 ms.

As an effect of the cell retransmissions that take place (which means additional traffic), the interference in the system increases and the traffic load that could have been carried decreases.

8.4.2.4 DCA with Asymmetric Traffic

Operating DCA with asymmetric traffic is not a capacity enhancement technique. It is interesting, however that the DCA technique performs slightly better for a realistic asymmetric traffic distribution than for a symmetric distribution, whereby the resulting overall traffic load is the same for both cases. Considering the traffic model, the offered traffic is determined by a T_{on} time and a T_{off} time. The mean T_{on} for a connection is 10 s. Adjusting the mean T_{off} time of the connection the traffic offer λ_{ABR} (RBS) of an RBS can be determined.

$$\lambda_{ABR}(\text{RBS}) = \sum_{i=1}^{n_{RNT}} \frac{T_{on_i}}{T_{off} + T_{on_i}} \cdot \text{MCR}_i \tag{8.10}$$

where MCR_i is the mean cell rate of a single connection i and n_{RNT} is the number of RNTs present in one radio cell (here equal to 48). The T_{off} calculated here is valid for

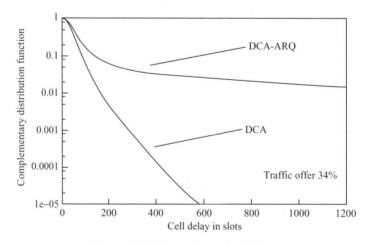

Figure 8.11   Comparison of cell delay

RNTs on the uplink and the downlink for a ratio of 1:1. To adjust an asymmetric load of, for example, 1:4 for uplink to downlink, where there is an equal number of terminals operating on the uplink and the downlink, the following calculation can be done:

$$\lambda_{\text{ABR}}(\text{RBS}) = 24 \cdot \frac{T_{\text{on,DL}}}{T_{\text{off,DL}} + T_{\text{on,DL}}} \cdot \text{MCR}_{\text{DL}} + 24 \cdot \frac{T_{\text{on,UL}}}{T_{\text{off,UL}} + T_{\text{on,UL}}} \cdot \text{MCR}_{\text{UL}} \qquad (8.11)$$

Considering $T_{\text{on,DL}} = T_{\text{on,UL}} = T_{\text{on}}$ and $\text{MCR}_{\text{DL}} = \text{MCR}_{\text{UL}} = \text{MCR}$ the asymmetry condition is

$$\frac{1}{4} = \frac{T_{\text{on}} + T_{\text{off,DL}}}{T_{\text{on}} + T_{\text{off,UL}}} \quad \text{which leads to} \quad T_{\text{off,UL}} = 3T_{\text{on}} + 4T_{\text{off,DL}} \qquad (8.12)$$

Thus, for the downlink the capacity can be determined by the $T_{\text{off, DL}}$ time derived indirectly from Equation (8.13)

$$\lambda_{\text{ABR}}(\text{RBS}) = T_{\text{on}}\left(\frac{30}{T_{\text{on}} + T_{\text{off,DL}}}\right) \cdot \text{MCR} \qquad (8.13)$$

and similarly for the uplink from Equation (8.14)

$$\lambda_{\text{ABR}}(\text{RBS}) = T_{\text{on}}\left(\frac{120}{T_{\text{on}} + T_{\text{off,UL}}}\right) \cdot \text{MCR} \qquad (8.14)$$

The traffic model can be understood as if 30 RNTs (half of them, 15, in downlink) with a high traffic load or 120 RNTs (60 in uplink) with a low traffic load would generate the same traffic at an RBS as 48 RNTs (thereof 24 in each direction, downlink and uplink) operating at a medium traffic load. The asymmetry has been implemented in the system in order to evaluate the DCA behaviour with respect to such an asymmetric traffic. The results are presented in the final comparison of all the investigated technologies at the end of this section.

8.4.2.5 DCA with Power Control

The intention behind power control is to overcome the so-called 'near–far' problem. The strong signal received at the cell site from a near-in RNT will mask over the weaker signal from a far-out RNT if both transmit with the same power. Applying power control, the transmitting powers of each station can be adjusted so that the received signal power at the radio cell site is the same from all RNTs and vice versa. There are two different methods of achieving power control. A primitive method is the adaptation of the transmitting power during installation of the system. The adaptation is calculated on the basis of the path loss model. A more sophisticated adaptive power control method can be accomplished by regulating the current transmitting power of each transmitting station according to the current C/I level of its respective connection. For the DCA power control simulation the simpler method was used. Considering the stationary nature of the RNTs in a FWA network scenario this strategy seems appropriate.

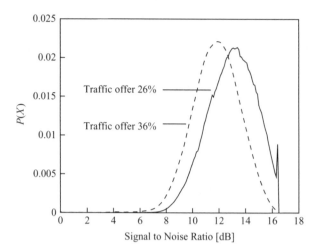

Figure 8.12 SNR in the power controlled system

Figure 8.12 shows the SNR curves for the uplink from RNT to RBS at two different traffic loads. Once again it can be seen that increasing the capacity required at the RBS, the SNR decreases. The employment of power control can be seen in the progression of the curve for an RBS capacity of 26%.

The highest possible SNR in the system is now 16.5 dB. This value can be explained by the following argument. The required received power level at each station is set to −78 dBm. Considering a background noise level of −97.5 dBm (see the radio parameters) and an additional fading margin of 3 dB exactly this SNR value of maximally 16.5 dB results for the radio cell in the case that no interference is present on the channel.

Comparing this value to the values in Figure 8.9 this maximum SNR is quite low. However, the adjusted received signal strength guarantees that the level of interference in the system is suppressed. Reducing the interference level in the radio cell has two effects. The amount of traffic that can be carried within a radio cell rises. Secondly, the PER is reduced significantly. Comparing the PER for the simple DCA scheme with the power controlled scheme, there is an almost three-fold advantage when the PER for the former is 0.05 and the PER for just DCA is 0.15 for a system operating at 34% traffic load. Plotting the PER against the radio cell radius, another interesting effect can be observed.

As can be seen in Figure 8.13 the PER is almost constant over the distance of an RNT from its RBS. This observation is conform to the purpose of the power control technology. Also noticeable is that the PER increases for the RNTs located at a distance of more than 1800 m from the RBS. The PER increases for these terminals because their received transmission power strength is below that of the predetermined level of −78 dBm. As a result these RNTs have a lower SNR at the RBS than the nearer RNTs.

The overall interference in the system has been suppressed at the cost that the far RNTs may have transmission errors. Also shown in Figure 8.13 is an imbalance between the PER for the downlink when compared to the uplink. The imbalance is due to the fact that the resource allocating entity is in the RBS. For transmissions on the uplink the RBS selects those resource units which are suitable for itself. However, while transmitting on the downlink the RBS selects resource units which may not be optimal from the point of view of the RNTs. Figure 8.13(b) shows how the PER increases in this case.

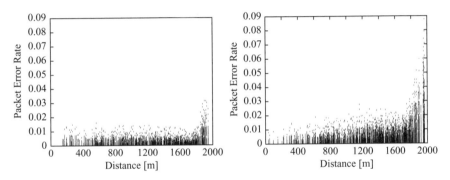

Figure 8.13 Mean PER for the uplink (left plot), mean PER for the downlink (right plot)

8.4.2.6 DCA with Directional Subscriber Antenna

Until now all simulations were performed with omni-directional antennas at the RNTs and the RBSs. Mitigation of interference was achieved by using power control. A further and more efficient reduction of interference is achieved if directional narrowbeam antennas are placed at the RNT sites.

For the simulations the antenna characteristics were selected according to the values contained in Table 8.5. Beneath the reduction of transmission power due to the high antenna gain of 29 dBi, the main signal component will be transmitted to the RBS and is not emitted in other directions.

The installation of narrowbeam antennas therefore leads to a reduction of interference, thereby increasing the SNR in the system. Raising the signal strength should culminate in a higher capacity utilization of the RBS and a reduced PER.

Figure 8.14 shows the comparison of the SNR value distribution between the simple DCA scenario and network employing directional antennas at the RNTs. The DCA simulation referenced here was simulated with a traffic capacity of 32 % while the simulation with directional antennas was performed with 42 % traffic load. Both systems showed similar performance of rejected and aborted connections at these traffic loads. The afore mentioned SNR improvement between the two schemes is also showed in Figure 8.14.

Due to a significantly lower number of possible interferes involved, the scattering of the SNR values is very high. Measuring a mean SNR for simple DCA at 15 dB and a mean SNR of 21 dB for the enhanced technique, an increase of 6 dB sets in.

Considering the results of the power control scheme, an even greater capacity enhancement should be achievable if power control is applied to the DCA scenario with narrowbeam antennas. This optimized scheme will be discussed later.

8.4.2.7 DCA with Sectored Cells

Sectoring at the RBS site allows a further reduction of transmission power and therefore also a suppression of overall system interference. The installation of high-gain antennas at both the RNT and RBS that means coupling sectored RBS antennas and narrowbeam RNT antennas mitigates the interference experienced by stations located beyond the opening angles of these antennas.

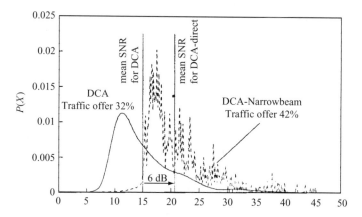

Figure 8.14   SNR for narrowbeam RNT antennas

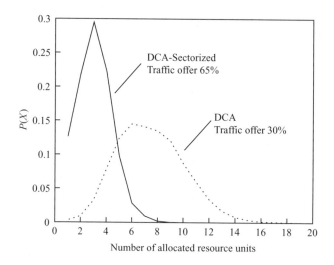

Figure 8.15   Allocated resource units

Sectoring is performed by splitting a radio cell into multiple pie shaped wedges. In the simulator each sector is considered as an independent radio cell with its own self-organizing RBS. Hence, from a technical point of view, the number of somehow virtual RBSs in a hexagonal radio cell increases according to the number of sectors in this radio cell.

The error performance improves greatly for a DCA scheme with sectored antennas. The mean PER stays at 0 in the sectored system for traffic offers up to 25%. Note that the traffic offer has to be considered at each virtual RBS, e.g. each sector of the radio cell. Compared to the simple DCA scheme with a capacity of 30% the sectored system is able to carry more than 150% traffic. This is possible, because the same frequencies which means the same part of the radio capacity can be reused within the same cell by different sectors. With the traffic model assumed for the simulation it was not possible to estimate the limit of the sectored FWA scenario.

8.4.3 Capacity Analysis of FCA and DCA for FWA Networks

The capacity analysis for the mentioned DCA schemes can be based on the major parameters mean SNR and the GoS accomplished in the systems.

8.4.3.1 Mean SNR

It is commonly expected that system performance increases with steadily rising SNR. However, this is not quite correct in the case for FWA networks.

Figure 8.16 gives an overview of the mean SNR values plotted against the simulated traffic loads for the different schemes. Here, it can be observed that power controlled DCA which can carry more traffic than simple DCA has a lower mean SNR. Similarly, FCA with a cluster size of 12 has a higher mean SNR than DCA although it can carry less traffic.

Apparently, the overall interference level in the system also plays a very important role in allocating capacity. Power controlled DCA has a low mean SNR but simultaneously also a low interference level in the system since the transmission power of the stations is reduced with respect to the other schemes. The FCA schemes lose capacity because the interference in the co-channel cells is too high. Striking is the extreme increase in SNR through the use of narrowbeam antennas at the RNT station sites, which is a result of the highly directional antennas emitting only 2.3 % of the power compared to an omni-directional antenna. Therefore, although the average transmission power is similar for all schemes, the mitigation of interference is by far highest in the scenarios with high-gain antennas.

8.4.3.2 Grade of Service

The second parameter for defining the capacity which can be offered by an RBS is its GoS. The GoS is defined by

$$\text{GoS} = \frac{\text{nos. of rejected calls} + (10 \times \text{nos. of aborted calls})}{\text{total number of calls}} \tag{8.15}$$

The number of aborted calls (also known as dropped calls) is weighted with 10 as this would mean that a terminal loses its connection while transmitting, a situation which is highly intolerable. A system should be operated at a GoS below 1 %. Figure 8.17 gives an overview of all the different DCA schemes that were simulated and their respective GoS performance.

It can be seen that FCA performs worse when compared to the DCA schemes. The notable difference is a little surprising concerning the very low amount of traffic that can be carried. The main reasons for this poor performance are the high co-channel inter-ference levels that are present for small cluster sizes (≤ 7) or the little amount of available resources for large cluster sizes (≥ 12). Since the traffic model used for simulations generates a highly dynamic behaviour, call blocking becomes a severe problem for the FCA schemes which are unable to flexibly react on varying traffic conditions.

Interesting is also the observation that DCA performs slightly better than DCA with ARQ. This effect is due to the higher interference caused by the retransmissions of the

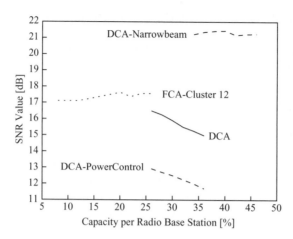

Figure 8.16   Mean SNR values

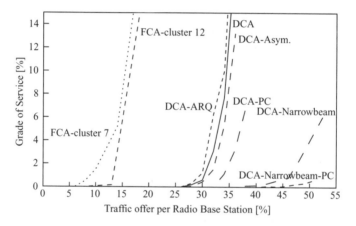

Figure 8.17   Comparison of GOS

ARQ protocol. A considerable raise in capacity is achieved if power control is applied. The power controlled DCA system can carry between 5 % and 6 % more traffic load at a comparable GoS. The application of narrowbeam antennas at the RNTs improves capacity by 14 % against the conventional DCA scheme. Considering the raise caused by the last two technologies it can be expected that a power controlled DCA scheme with narrowbeam antennas should lead to a capacity increase of around 20 % against just DCA. The highest gain in capacity per RBS is achieved by sectoring of an RBS and the employment of narrowbeam RNT antennas. This FWA network optimized technology leads to a gain of more than 80 % compared to the simple DCA scheme. If within the sectored scheme RNTs are allowed to measure the interference situation in resource unit an additional gain of around 2 % should be registered.

List of Abbreviations

ABR	Available Bit Rate
ACI	Adjacent Channel Interference
ARQ	Automatic Repeat reQuest
ATM	Asynchronous Transfer Mode
C/I	Carrier to Interference ratio
CBR	Constant Bit Rate
CCI	Co-Channel Interference
CD	Code Division
CDMA	Code Division Multiple Access
CLR	Cell Loss Rate
CRE	Cell Reference Event
CS	Channel Segregation
CTD	Cell Transfer Delay
DAVIC	Digital Audio Video Council
DCA	Dynamic Channel Allocation
DCS	Dynamic Channel Selection
DECT	Digital Enhanced Cordless Telephone
FCA	Fixed Channel Allocation
FD	Frequency Division
FDMA	Frequency Division Multiple Access
FEC	Forward Error Correction
FWA	Fixed Wireless Access
GaAs	Gallium Arsenid
GoS	Grade Of Service
HCA	Hybrid Channel Allocation
ICI	Inter-Channel Interference
LoS	Line of Sight
LMDS	Local Multipoint Distribution System
MADCAT	Mobile ATM Dynamic Channel Allocation simulaTor
MAC	Medium Access Control
MCR	Mean Cell Rate
PDU	Packet Data Unit
PER	Packet Error Ratio
QAM	Quadrature Amplitude Modulation
QoS	Quality of Service
QPSK	Quaternary Phase Shift Keying
RACH	Random Access CHannel
RBS	Radio Base Station
RMS	Root Mean Square
RNT	Radio Network Terminal
RSC	Reed-Solomon Code
SNR	Signal-to-Noise Ratio
TD	Time Division
TDMA	Time Division Multiple Access
UBR	Unspecified Bit Rate

VBR	Variable Bit Rate
VoD	Video on Demand
WATM	Wireless ATM
WLL	Wireless Local Loop

References

[1] J. B. Andersen, T. S. Rappaport and S. Yoshida, *Propagation Measurements and Models for Wireless Communication Channels. IEEE Commun. Mag.*, vol. 44, pp. 163–171, Jan. 1995.

[2] DAVIC (1999) *DAVIC 1.1 Specification*. Technical Report, Digital Audio-Visual Council, Geneva, available at *ftp://ftp.davic.org/Davic/Pub/Spec1 1/prt08 11.doc*.

[3] I. Forkel, R. Easo and S. Mangold, 'Computer Simulations of Performance Enhancing Methods in ATM based Fixed Wireless Access Networks,' in *Proceedings of SPECTS 2000*, Vancouver, B C, Canada, June 2000.

[4] U. Karthaus, R. Noë and A. Gräser, 'High Resolution Channel Impulse Response Measurements for Radio in the Local Loop,' in *ACTS Mobile Communications Summit*, Aalborg, Oct. 1997.

[5] I. Katzela and M. Naghshineh, 'Channel Assignment Schemes for Cellular Mobile Telecommunication Systems: A Comprehensive Survey,' *IEEE Personal Commun.*, pp. 10–30, June 1996.

[6] A. Krämling and M. Scheibenbogen, 'Influence of Channel Allocation on the System Capacity of Wireless ATM Networks,' in *2nd International Workshop on Wireless Mobile ATM Implementation (wmATM'99)*, June 1999.

[7] S. Mangold and I. Forkel, 'Optimal DLC Protocol Configuration for Realistic Broadband Fixed Wireless Access Networks based on ATM,' in *The Second International Conference on ATM (ICATM '99)*, Colmar, France, June 1999.

[8] A. R. Noerpel and Y.-B. Lin, 'Wireless Local Loop: Architecture, Technologies and Services,' *IEEE Personal Commun*, pp. 74–80, June 1998.

[9] NORTEL NETWORKS Local Multipoint Distribution Service (LMDS) Tutorial. available at: http://www.webforum.com/nortel4, April 1999.

[10] T. S. Rappaport and V. Fung, 'Simulation of Bit Error Performance of FSK, BPSK and Pi/4-DQPSK in Flat Fading Indoor Radio Channels Using a Measurement-Based Channel Model,' *IEEE Trans. Vehicular Technology*, vol. 40, pp. 731–740, Nov. 1991.

[11] Regulierungsbehörde für Telekommunikation und Post, RegTP Funkanbindung von Teilnehmeranschlüssen mittels Punkt-zu-Mehrpunkt-Richtfunk (WLL-PMP-Rifu), available at: http://www.regtp.de/Aktuelles/aktuelle1.htm, 1999.

[12] Video Information Provider Consulting, VIPC Case Study: Stanford Telecom and LMDS, available at: http://www.dnai.com/~desmith/stanfordtel.html, Oct. 1996.

9

WLL as an Interferer

Miroslav Dukic

9.1 Introduction

In the past three decades there were big social, economic and technological changes in the world. Scientific and technological advancement, informing of wide circle of people and their inclusion in the learning process are the main characteristics of this period. In such conditions, the need came up for exchanging different kind of messages, i.e. transmission of different kind of signals, governing the extremely fast development of the telecommunications systems. With utmost certainty it can be said that there are very few human activities that experienced such basic, qualitative and quantitative changes in their development as modern telecommunications.

One of the characteristic examples of the fast development and wide usage of new technologies are professional radio systems. Terrestrial microwave links and corresponding satellite telecommunication systems are the most widely used professional radiosystems. Together with intercontinental cable links, they are today the grounds for national and worldwide telecommunication systems.

The basic problem in the current state of development of modern professional radio systems is their coexistence. The fact is that the technologically available bands and satellites orbits inherent and restricted resources show that their usage must be rational and that they are the main factor in solving the coexistence problem between old systems and newly developed PCS (*Personal Communication Systems*). As a consequence of limited available frequency bands different professional radio systems operate in the same, or near, frequency bands, which is very disadvantageous regarding the coexistence.

On the other side, the global need for telephone network access is driven by pent-up demand for existing telecommunications services, by economic pressure to expand a region or nation's access to telecommunications, and by the impacts of deregulation. The deployment of central office switches and trunk capacity, however, represents the easiest part of expanding a nation or region's telephone infrastructure when compared to the effort required to provide network access to each subscriber.

Modern digital techniques with high information capacity and efficient spectrum utilisation are revolutionizing the capability of cellular communications networks to provide new services to subscribers. As an example, today's mobile communication systems are primarily designed to provide cost efficient wide area coverage for users with moderate bandwidth demands. Extending PSTN (*Public Switched Telephone Network*) to the mobile community has been the main driving force in the evolution we have seen so far.

In the last few years there is a significant world trend of extending the existing and installing new advanced telephone systems using the same realization principles as those used in cellular networks. These systems are known as WLL (*Wireless Local Loop*), WiLL (*Wireless in Local Loop*), RiLL (*Radio in Local Loop*) or FRA (*Fixed Radio Access*). Driven by many advances in radio technology and manufacturing process commonly associated with the mobile cellular industry, WLL has recently become an economically attractive alternative to traditional wired outside plant. For operators of telephony networks, outside plant often constitutes the major capital expense, and the choice of WLL can impact over half of their typical investment expenses. The cost advantage that WLL offers over traditional wire fixed line can thus have a major impact on a service provider's bottom line. Therefore, even though the mobile communication systems and WLL systems may appear to be similar, and sometimes even used interchangibly, the requirements are quite distinct.

Mobile cellular networks, by their very nature, must spend considerable processing resources on the tasks of tracking the spatial location of users, and allowing their dispersion to undergo rapid dynamic change. With fixed subscribers, such tasks are not needed. The location of subscribers does not undergo a dynamic change. Since the direction of a subscriber relative to a serving base station is fixed, WLL antennas may exploit the benefits of directionality. The best of WLL technologies and products can therefore provide significantly higher subscriber densities, higher call capacity and better quality of service than their mobile counterparts.

To be a true commercial substitute for wireline, WLL systems seek to provide transparency. WLL is most attractive when it behaves in a similar manner to high quality wireline telephony, but at considerably lower cost. The best of WLL technologies and products available today achieve excellent transparency, both for analogue as well as digital telephone service. Indeed, the highest compliment that can be paid to a WLL product is for a typical end user not to be able to detect that a call is using a WLL line.

One of the biggest problems in the design of modern WLL systems is the choice of frequency bands for their operations. A wireless communication system has to recognize that the frequency bands available will always be limited. The key focus has to be efficient use and re-use of the spectrum. The use and re-use of the spectrum is considered by many factors including:

- Symbol rate.
- Signalling overhead.
- Modulation efficiency.
- WLL cell-radius.
- Choice of multiple access.
- Possible interference reduction techniques.
- Spatial diversity and space-time processing.
- Electromagnetic coexistence.

Since WLL operates as a public outdoor radio technology, to reduce interference it must operate only in licensed radio bands. The exact frequencies under which WLL systems operate are therefore controlled by national, regional and international regulatory bodies. Public telephone service providers seeking to operate WLL systems generally must apply for radio spectrum in the locations in which they wish to operate. Common operating frequencies of modern WLL systems are in the 1.9 GHz and 3.4 GHz bands. In the near future the usage of new frequency bands, up to 30 GHz, is planned. For example, some WLL radio technologies such as DECT (*Digital Enhanced Cordless Telecommunications*) offer advanced radio techniques such as dynamic channel selection to provide a high level of coexistence and excellent spectrum efficiency, relate to existing professional radio systems.

With many different WLL radio technologies on the market and systems with different qualities of service and different levels of transparency, there is hesitancy by some operators to adopt WLL. Purpose-built WLL systems, which already have good transparency, are pressured now to standardize and align their operations and management systems with the rest of the network. Most significantly, the demand of operators that WLL systems support ever-higher data rates requires the vendors to continually evolve their systems.

As modern technologies such as V.90 modem, xDSL (*Digital Subcsriber Lines*), and cable modems are deployed, WLL systems are pressured to match their capabilities. Only systems using the most modern digital radio technologies, such as DECT, TDMA (*Time Division Multiple Access*) and CDMA (*Code Division Multiple Access*) technologies are likely to maintain significant WLL market share. For even in the least developed areas, urban and rural, there is both a need and a demand for modern data services such as Internet access at ever increasing data rates.

The requirement to support continually higher data rates suggests that the introduction of packet technologies over WLL radio interfaces will become a commonplace in the next few years. Instead of connecting only to traditional circuit switches, we are likely to see WLL systems directly interface to IP routers as well. It is the ability of packet technology to increase the sharing of radio resources, which drive the interest in applying packet technology to WLL. With increasing deregulation, traditional as well as new operators may seek to provide both circuit switched telephony services as well as packet switching for services such as Internet access.

Radio technologies that can dynamically adapt to asymmetry have a distinct advantage over those that do not. In particular, if the duplexing of two-way communications is achieved by means of TDD (*Time Division Duplex*), it is significantly easier to adjust to asymmetry in real time than with FDD (*Frequency Division Duplex*).

In the near future, WLL systems are likely to continually incorporate various new technological advances such as smart antenna technology, the dynamic alternation of the shape of electromagnetic propagation, to improve performance. A number of methods for this have been demonstrated. For WLL systems, greater capacity will be achieved by reduced interference and more efficient use of radiated power.

At the end of this introduction, it should be stressed that the development of the future WLL systems depends mainly on the choice of the radio interface technology, that is to say, on solving the problem of coexistence with microwave links.

The whole material presented in this chapter can be, generally, divided into three equal parts.

The first part of the presentation is concerned with technical[1]–technological aspects of using the modern WLL systems, and the problems of their coexistence with the present microwave links, using the same frequency bands. To achieve the electromagnetic coexistence of these systems, the chosen technology for the WLL systems radio interface must inherently be a small source of interference, conforming to all conditions regarding capacity and services for modern WLL systems and that is inherently robust (immune) to exterior interference. At the present technological level SSDS WCDMA (*Spread Spectrum Direct Sequence, Wide-band Code Division Multipe Access*) is an optimal solution.

The second part of the presentation is concerned with the presumed operating scenario of WLL and microwave FDM/FM (*Frequency Division Multiplex, Frequency Modulation*), that is, digital microwave link (DML).

The last part of the presentation in this chapter is concerned with the quanitative interference analyses that the WLL system using WCDMA technology originates in fixed service microwave link. The derived results show that, in specific conditions, the coexistence of these systems is possible.

9.2   System Overview of WLL as an Interferer

9.2.1   Scenarios of WLL Systems Implementation

Many different scenarios can be applied for the deployment of WLL, ranging from high-density urban areas through the suburbs and rural communities:

- *Existing* operator—*serving a new area*: The use of WLL systems in these situations allows investment in telephone structure to follow the demand of new services and subscribers.
- *Existing operator—rural area*: In rural area most of subscribers are typically clustered in small villages clusters at distances up to 30 km from the exchange.
- *Existing operator—expanding capacity*: New demands for services are typical for un-developed areas, urbane or rural, in underdeveloped countries,
- *New operator*: The main goal in this case is to provide services as rapidly and cost-effectively as possible. This scenario is becoming very important for the developed countries.

The advantages of using WLL systems are becoming known to an increasing number of service providers. The advantages are particularly valuable in areas where the demand for services is increasing and the deregulation of the telephone industry is introducing competition into markets and technology segments that were once monopolies.

Wireless technology offers numerous advantages over copper wire local loops that have been proven in field tests and deployed systems around the world. The basic advantages of the application of WLL systems are summarized as follows:

- Avoiding the extremely high investments in building the fixed telephony infrastructure.
- For new operators and existing operators with tight constraints on available investment capital, it is the incremental, modular nature of WLL and its speed of deployment that are key attractions. WLL can generally bring a return on investment much faster than wireline deployment because it can be deployed faster. WLL also allows the investment to be made in smaller increments, tracking demand and return on investment.
- Low incremental cost for adding users once the base stations.
- Building the WLL telephone network requires significantly less time than building the fixed telephony network.

- Interconnection with PSTN is simple to establish.
- Future expansion is very simple.
- Network maintenance costs are lower.
- Indifference to topography and distance.
- The WLL system should allow encryption of the radio interface and fraud prevention capabilities.
- Modern WLL technology shares some aspects of the common architecture of mobile systems—cellular technology, sectorization, frequency re-use, low power, etc.

Operators already are aware that a successful WLL technology must meet standards in the following five areas:

- Dropped calls and fades.
- Interference due to crosstalk.
- Privacy.
- Blocking rates.
- Voice quality.
- High degrees of compatibility and transparency of performance, operation, billing and network management as in fixed telephone services.
- Electromagnetic coexistence.

The WLL system may provide the following general services, at a minimum:

- *Voice:* The system may provide full switched toll grade quality voice service. The voice quality may be telephone toll grade or better and there may be no delays in speech that are perceptible to the user. The voice user is not expected to change any of their infrastructure interfaces. The normal telephone connection may be provided by means of the LMDS (*Local to Multipoint Distributed System*) local interface unit. The system must also provide all typical custom calling features as expected in normal delivery of a competitive wire based telecommunications service.
- *Low Speed Data:* The system may be able to provide data at the rates up to 9.6 kbps on a transparent. The system may handle all data protocols necessary in a transparent fashion. The network may allow local access to value added networks from the local access point. The low speed data may be provided for over a standard voice circuit from the users premises as if there were no special requirement. The system may also be capable of support all Group 3 fax services.
- *Medium Speed Data:* The network may be able to handle medium speed data ranging up to 64 kbps. The interfaces for such data may be the value added network local nodes. The medium speed data may be provided for over a standard voice circuit from the users premises as if there were no special requirement. The interconnection for 64 kbps may also be ISDN (*Integrated Service Digital Network*) compatible.
- *High Speed Data:* Data rates at 2 Mbps may also be provided on an as needed basis and a dedicated basis.
- *Video:* The network may be able to provide the user with an access to analogue and digitized video services. This may also enable the provisioning of interactive video services.

On the other hand, the disadvantages of the WLL systems can be stated as follows:

- In developing countries, where the potential market for WLL exists and where continuous supply of power may not be so certain, the base stations.
- User's equipment will need power supply locally and in the event of the power failure the service to a user or a group of users will be lost.
- The technology has still not stabilized and, as a result, the performance of the present day wireless communication services is not of top quality with frequent dropping of calls, unsatisfactory levels of noise, etc.
- With obsolescence of the equipment installed, there being fast development in this area, replacement of the same at considerable expenses may have to be done by the service provider.
- In the absence of sufficient technically skilled personnel, particularly in developing countries, the repair or replacement of base stations and user's units will cause problems.

Here, it should be stressed that foregoing advantages and weaknesses of WLL systems depend mostly on the chosen radio interface technology for WLL system and solving the coexistence problem with existing radio systems.

9.2.2   Technology of WLL Systems

The WLL revolution is underway. WLL suppliers and operators are flocking to emerging markets, using whatever available wireless and line interface technologies are at hand to achieve fast time to market. Since there are no definitive WLL standards, vendors are faced with a bewildering choice of fixed-access, mobile, and digital cordless technologies. Ultimately the appropriate protocol technology will depend on an array of application considerations, such as size and population density of the geographic area (rural versus urban) and the service needs of the subscriber base (residential versus business; PSTN versus data access). In fact, there are many good reasons why different wireless technologies will serve some applications better than others.

The challenge for WLL vendors is to identify the optimal wireless protocol for their unique application needs, then reduce cost per subscriber and deliver integrated solutions to the marketplace. WLL will be implemented across five categories of wireless technology. They are digital cellular, analogue cellular, personal communications network, personal communications service, DECT (Digital European/Enhanced Cordless Telecommunication), and proprietary implementations. Each of these technologies has a mix of strengths and weaknesses for WLL applications.

In the following text we shall take a look at the characteristics of the mentioned technologies regarding the coexistence of the new WLL system and the existing professional radio systems.

9.2.2.1   Analogue cellular

Given its wide availability resulting from serving high-mobility markets, there is significant momentum to use analogue cellular for WLL. There are currently three main analogue cellular system types operating in the world: AMPS (*Advanced Mobile Phone System*), NMT (*Nordic Mobile Telephone*), and TACS (*Total Access Communications System*). AMPS dominate the analogue cellular market with 69 % of subscribers, TACS has 23 % and NMT has only 8 % of the global subscribers.

As a WLL platform, analogue cellular has some limitations in regards to capacity and functionality. Due to widespread deployment, analogue cellular systems are expected to be a major wireless platform for WLL, at least in the short term. Given its characteristics, analogue cellular is best suited to serve low-density to medium-density. Analogue cellular is forecasted to account for 19 % of the WLL subscribers in the year 2001.

With regard to the coexistence, the choice of analogue WLL system is a bad solution. This type of WLL system is the source of strong interference in the present radio systems, and, at the same time, WLL systems themselves are susceptible to the exterior interference.

9.2.2.2 Digital Cellular

These systems have seen rapid growth and are expected to outpace analogue cellular over the next few years. Major worldwide digital cellular standards include GSM (*Global System for Mobile Communications*), hybrid solution of TDMA and FDMA, and CDMA. GSM dominates the digital cellular market with 71 % of subscribers.

Digital cellular is expected to play an important role in providing WLL. Like analogue cellular, digital cellular has the benefit of wide availability. Digital cellular can support higher capacity subscribers than analogue cellular, and it offers functionality, that is better suited to emulate capabilities of advanced wireline networks.

It is very significant that the digital WLL systems are a relatively weak source of interference, which facilities to a considerable extent conditions of electromagnetic compatibility. Its disadvantage is that it is not as scalable as analogue cellular.

It is forecasted that approximately one-third of the installed WLLs will use digital cellular technology in the year 2001. Although GSM currently dominates mobile digital cellular, there has been little activity in using GSM as a WLL platform. Since GSM's architecture was designed to handle international roaming, it carries a large amount of overhead that makes it unwieldy and costly for WLL applications. In spite of these limitations, it is likely that GSM WLL products will be developed over the next few years.

CDMA appears to be the standard best suited for WLL applications. CDMA employs a spread-spectrum modulation technique in which a wide range of frequency is used for transmission and the system's low-power signal is spread across wide-frequency bands. It offers higher capacity than the other digital standards (10 to 15 times greater than analogue cellular), relatively high-quality voice, and a high level of privacy. The main disadvantage of CDMA is that it is only now beginning to be deployed on a wide scale.

9.2.2.3 PCS

PCS (Personal Communication System) incorporates elements of digital cellular and cordless standards as well as newly developed *radio-frequency* (RF) protocols. Its purpose is to offer low-mobility wireless service using low-power antennas and lightweight, inexpensive handsets. PCS is primarily seen as a city communications system with far less range than cellular.

PCS is a broad range of individualized telecommunications services that let people or devices communicate regardless of where they are. Some of the services include personal numbers assigned to individuals rather than telephones, call completion regardless of locations, calls to the PCS customer that can be paid by either the caller or the receiver, and call-management services that give the called party greater control over incoming calls.

At this time, it is not clear which standards, if any, will dominate the WLL portion of PCS. The candidate standards are CDMA, TDMA, GSM, PACS (*Personal Access Communication Systems*), omnipoint CDMA, TDMA, upbanded CDMA, PHS (*Personal Handyphone System*), and DCT-U (*Digital Cordless Telephone United States*). These standards will probably be used in combination to provide both WLL and high-mobility wireless services. PCS has the advantage of being designed specifically to provide WLL by public wireless operators.

9.2.2.4 DECT

DECT was originally developed to provide wireless access within a residence or business between a base station and a handset. Since the base station is still hard-wired to the PSTN, this is not considered WLL. For the purposes of this study, DECT is considered WLL when a public network operator provides wireless service directly to the user via this technology.

Although DECT does not appear to be ideally suited for WLL in rural or low-density applications, it has some significant advantages in medium-density to high-density areas. Cordless telephony has advantages in terms of scalability and functionality. As compared to cellular technology, DECT is capable of carrying higher levels of traffic, provides better voice quality, and can transmit data at higher rates. The microcell architecture of DECT allows it to be deployed in smaller increments that more closely match the subscriber demand, with reduced initial capital requirements.

9.2.2.5 Background and standardization of radio interface for IMT-2000 system

ITU-R TG 8/1 at the Helsinki meeting (November 1999) approved a comprehensive set of terrestrial and satellite radio interface specifications for IMT-2000. The terrestrial component encompasses the following five different technologies:

- UTRA (*Universal Terrestrial Radio Access*) FDD (WCDMA) specifications are being developed within the 3GPP. This radio access scheme is direct-sequence CDMA with information spread over approximately a 5 MHz bandwidth with a chip rate of 3.84 Mchps.

  The radio interface carries a wide range of services to support both circuit-switched services and packet-switched services.
- CDMA 2000 specifications are currently developed within the 3GPP2 for the multi-carrier version of IMT-2000. It is a wide-band spread spectrum radio interface with CDMA technology.

  The physical layer supports RF channel bandwiths of $N \times 1.25$ MHz, where N is the spreading rate number.
- UTRA TDD and TD-SCDMA specifications are currently developed within the 3GPP. UTRA TDD has been developed with the UTRA FDD part by harmonizing important parameters of the physical layer and specifying a common set of protocols in the higher layers. TD-SCDMA has significant commonality with the UTRA TDD. Specifications include capabilities for the introduction of TD-SCDMA properties into a joint concept. The radio access scheme is DS-CDMA.

  UTRA TDD spreads information over approximately a 5 MHz bandwidth and has a chip rate of 3.84 Mchps. TD-SCDMA spreads information over approximately 1.6 MHz bandwidth and has a chip rate of 1.28 Mchps.

- UWC-136 specifications are developed with inputs from the Universal Wireless Communications Consortium. This radio interface has been developed with the objective of maximum commonality with GSM/GPRS. It maintains the TDMA community's philosophy of evolution from 1G to 3G systems.

  A three-component strategy enables the 136 technology to evolve towards 3G by enhancing the voice/data capabilities of the 30 kHz channels (designated as 136+), adding a 200 kHz carrier component for high speed data (384 kbps) for accommodating high mobility (designated as 136HS Outdoor), and adding a 1.6 MHz carrier component for data up to 2 Mbps in low mobility applications (designated as 136HS Indoor).

- DECT specifications are defined by a set of ETSI standards. The standard specifies a TDMA radio interface with TDD duplexing. The radio frequency bit rates for the modulation schemes are 1.152 Mbps, 2.304 Mbps and 3.456 Mbps.

  The standard supports symmetric and asymmetric connections, connection oriented and connectionless data transport, and variable bit rates up to 2.88 Mbps per carrier.

9.2.2.6 Multiple Access Technologies

The existing WLL systems use both conventional techniques of multiple access, FDMA and TDMA. However, these multiple access techniques have serious following drawbacks [12]:

- Necessity of providing the new frequency bands.
- Capacity of these systems is frequency and time limited.
- They are a significant source of interference in existing microwave links.

Using the technology and the experience in developing the third generation of the cellular networks that are using the SSDS-CDMA (*Spread Spectrum Direct Sequence Code Division Multiple Access*), the new generation of the WLL systems has been developed [10, 11, 13, 14, 15, 19, 20]. The main characteristics of this new generation of WLL systems are:

- Systems are inherently resistant to interference, with simultaneous time, frequency and space diversity.
- Frequency reuse factor is 1.
- WLL systems with broadband SSDS-CDMA belong to the class of low probability of interception systems.
- System capacity is limited only with expectable internal interference, which is produced by subscribers. Comparing with the conventional FDMA or TDMA WLL systems, system capacity could be increased by up to 15 times depending on the operating conditions.
- System concept allows easy connecting to the ISDN, PBX, PSTN or the existing resident cordless telephones.
- Improved voice privacy is built-in characteristic of SSDS systems.

9.2.2.7 WCDMA Basic Characteristics

A spread spectrum CDMA scheme is one in which the transmitted signal is spread over a wide frequency band, much wider than the minimum bandwidth required to transmit the information being sent. It employs a waveform that for all purposes appears random to anyone but the intended receiver of the transmitter waveform. Actually, for ease of both generation and synchronization by the receiver, the waveform is pseudorandom, but statistically it satisfies nearly the requirements of a truly random sequence. In the spread spectrum CDMA all users use the same bandwidth, but each transmitter is assigned a different code.

The important concept of WCDMA is the introduction of an intercell asynchronous operation and the pilot channel associated with each data channel. The pilot channel makes coherent detection possible on the reverse link. Furthermore, it makes it possible to adopt interference cancellation and adaptive antenna array techniques at a later date. It is well known that cell sectorization can increase link capacity significantly; the adaptive antenna array is viewed as adaptive cell sectorization and is very attractive. Other technical features of WCDMA are summarized below [2]:

- WCDMA support high bit rates, up to 2 Mbps. A variable spreading factor and multicode connections are supported.
- The chip rate of 3.84 Mchps used leads to a carrier bandwidth of approximately 5 MHz. DS-CDMA systems with a bandwidth of about 1 MHz, such as IS-95, are known as narrowband CDMA systems.
- The inherently wide carrier bandwidth of WCDMA has certain performance benefits, such as increased multipath diversity.
- WCDMA supports two basic modes: FDD and TDD.
  - FDD mode, with carrier separation of 5 MHz, are used for the uplink and downlink respectively, whereas in TDD only one 5 MHz is time-shared between uplink and downlink.
  - WCDMA system also for the unpaired spectrum allocations of the ITU for the IMT-2000 systems.
- WCDMA supports the operation of asynchronous base stations, so there is no need for a global time reference.
- WCDMA employs coherent detection on uplink and downlink based on the pilot symbols or common pilot. Coherent detection on the uplink will result in air overall increase of coverage and capacity on the uplink.
- The WCDMA air-interface has been crafted in such a way that advanced CDMA receiver concepts, such as multiuser detection and smart adaptive antennas, can be deployed to increase capacity and/or coverage.
- WCDMA is designed to be deployed in conjunction with GSM.
- Fast cell search under intercell *asynchronous* operation may be performed.
- Coherent spreading-code tracking.
- Fast transmit power control on both mobile-to-cell-site and cell-site-to-mobile links. Adaptive power control is used with minimum step size of up to 1 dB.
- Orthogonal multiple spreading factors in the forward link.
- Variable-rate transmission with blind rate detection.
- PN Sequences: Multirate codes, where the basic component is typically a Gold Code.
- Time Diversity (RAKE) is used in all systems.
- Bit Error Rates: designs vary from 10^{-3} to 10^{-5} for voice, 10^{-10} for data and 10^{-7} for video communications.
- Tolerable Dopplers: Up to 500 Hz are expected.
- Interference Cancellation.

9.3 Architecture of WLL as an Interferer

Modern WLL systems can be categorized in three following ways, especially including aspects of coexistence with fixed microwave links:

- *Physical implementation.* The categorization is based upon the way the WLL has been implemented—fully or partly wireless:
  — *Partly physical and partly wireless.* The connection to the user locality is physical—copper or fibre. Beyond the street crossing it is wireless. This can still be effectively used in congested areas.
  — *Fully wireless.* In this architecture the end-to-end link is wireless.
- *Usage categorization.*
  — *Fixed radio access.* In this case the user terminal is fixed with no mobility. Each end user communicates through base stations.
  — *Neighbourhood telephony.* In this case the user is able to move around in the house or in the immediate neighbourhood. Handover is possible within the same base stations but not beyond.
  — *Neighbourhood telephony with indoor base station.* In this case at the user end there is a PABX (*Private Automatic Branch Exchange*) interacting with the end user on one hand and base station on the other.
- *Technology based categorization.*
  — *Cordless telephony based.* For example such systems are DECT and PHS.
  — *Cellular based.* The cellular mobile communication technology is applied to provide local loop. Such systems are AMPS, NMT, GSM, etc.
  — *Point-to-point conection.*
  — *Satellite based.*

9.4 Problem Definition

Considering all above-mentioned characteristics of modern WLL systems, the results of interference analysis in FS-FDM/FM (*Fixed Service, FDM/FM*) and FS/DML (*Fixed Service Digital Microwave Link*) due to new WLL systems using SSDS-CDMA, are evaluated in this chapter. Effects of WLL systems are expressed through the Interference Noise Power at the FDM/FM receiver and BER (*Bit Error Rate*) at the DML receiver output, respectively.

As far as WLL are concerned, interference will be examined into FS-ML, even when WLL operates in the same frequency band. The results will be of great importance for the development of telecommunication systems in rural areas, because they will answer the question whether it is possible to develop telecommunication systems fast and economically. The main factors which contribute to the pressure for quick answers are:

- Technical advances.
- Tremendous economic and social pressure to expand telecommunications services.
- Deregulation.
- Coexistence with existing radio systems, especially microwave links.

9.5   Description of the Systems

9.5.1   System Layout

The block diagram of the system under consideration is shown in Figure 9.1. Block diagram is shown in two parts. In the upper part two kinds of fixed radio systems are shown, in which the interference influence from WLL systems is analysed. As the interference influence specification differs in analogue and digital fixed radio links, we have defined and analysed them separately.

The total interference originating from WLL system can be divided into:

- The interference from base stations, *forward* link.
- The interference from the WLL users, *reverse* link.

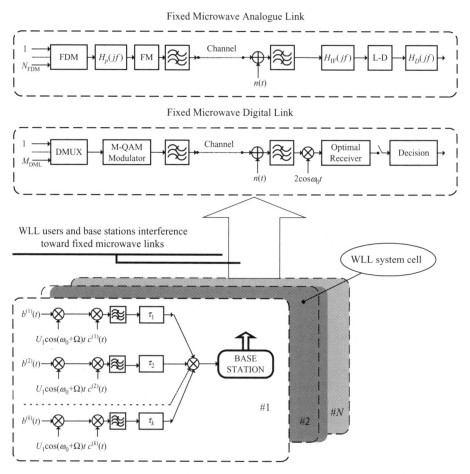

Figure 9.1   The system block diagram: DMUX—Digital Multiplexer, M_{DML}—Number of telephone channels in FS-DML, FDM—Frequency Division Multiplex, N_{FDM}—Number of telephone channels in FS-FDM/FM, FM—Frequency Modulation, $H_p(jf)$—Preemphasis transfer function, $H_{IF}(jf)$—IF filter, L-D—Limiter-discriminator, $H_D(jf)$—Deemphasis, K—Number of users per cell in WLL system

The considered WLL system is using FDD access, which means that *forward* and *reverse* links are isolated by wide an unused frequency bands. It is reasonable to assume that in dependence of spectral position one or the other direction of the link can be analysed independently regarding the interference on the fixed radio systems.

Unlike the cellular mobile telecommunication system, WLL systems have different initial assumptions, mainly with traffic and system topology. Because of the variety of WLL systems topologies, we have analysed only one rural model.

Typical system used as WLL in the rural environment is shown in Figures 9.2 and 9.3.

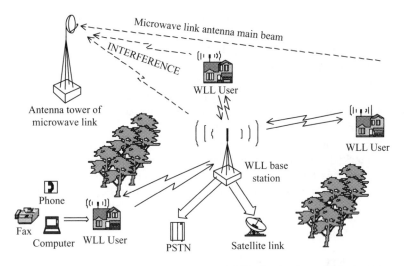

Figure 9.2 An example pf FSML amd WLL system spatial distribution in typical suburban or rural area

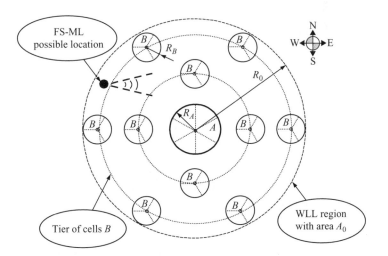

Figure 9.3 General Layout of WLL and FS-ML system

In the settings of the assumed circular territory with radius R_0 circular cells are distributed in the following way. In the centre there is a cell with radius R_A with K_A fixed subscribers. On the concentric circles with centre in the centre of the territory, are regularly distributed B type cells with radius R_B with K_B fixed subscribers. Cell A is larger than cell B and it represents central part of the region, usually the central part of inhabited settlement, while B-cells cover smaller settlements in the vicinity.

The number of tiers and the B-cell's size depends on the sub-urban or rural environment.

Base stations, that is, their antennas could be omnidirectional or directed in case of sectorization. Subscribers' antennas can be omnidirectional, but, as opposed to mobile systems, they can be directed to base station, which improve the quality of the link. It is because their position is fixed and known in advance.

Geometry of the system is defined in one pair of systems FS-ML—one WLL cell. Parameters which define current location of FS-ML, terminals and base stations in observed WLL system are given in Figure 9.4.

All antennas in the system are on specified heights. The adopted three-dimensional co-ordinate system (r, φ, z). The origin of the co-ordinate system could be chosen anywhere in the observed territory, with the zero height at the attitude of the surrounding terrain. It is assumed that the origin point of the co-ordinate system is at the location of the base station for the central cell of the observed WLL system.

The distances in the horizontal plane are $r_{x,y}$, while the real distances between the antennas are marked with $d_{x,y}$. Angles $\theta_{x,y}$ are azimuths of the antenna FS-ML axis relative to the antennas BS and MT in the system. Indices x, y in the specified dimensions are related to the corresponding locations in the system under observation, as shown in Figure 9.4.

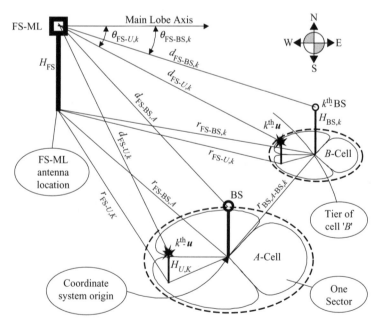

Figure 9.4   System geometry overview; HFS, HBS, k and HU, k are the antenna heights of FS-ML, the kth-U subscriber unit and the kth-BS base station in the WLL, respectively

All cells of the mobile and WLL system are partitioned in such a way that the main sector beams, in relation to the centre of the cell, are directed to 0°, 120°, and 240° (relative to the east, counterclockwise).

9.5.2 Cells and users distribution

The number of cells is defined by the area size, subscriber's density, and their spatial distribution. Assuming the uniform distribution of B-cells, Figure 9.4, maximum numbers of tiers and B-cells are $n_{T,\max}$ and $N_{B,\max}$, respectively, and given by M. Y. Dukic and M. Babovic [5]

$$n_{T,\max} = \frac{R_0 - R_B - R_A}{R_B} \tag{9.1}$$

$$N_{B,\max} = 3n_{T,\max}(n_{T,\max} + 1) \tag{9.2}$$

The probability density function of cells inside the WLL region of area A_0 is

$$dN_B = \frac{N_B}{A_0} r \, dr d\varphi, \tag{9.3}$$

while the number of users in the differential area, inside the cell of area A_C, is given by

$$dK_{Tr} = \frac{K_{Tr}}{A_C} r \, dr d\varphi, \tag{9.4}$$

where r and φ are polar coordinates.

9.5.3 Antenna Patterns Diagram

9.5.3.1 FS-ML Antenna Patterns

The FS-ML antenna radiation diagram is given by Y. R. Tsai and J. F. Chang [22]

$$G_{FS}(\theta) = \begin{cases} 32\,dBi, & 0° \leq \theta \leq 1°, \\ 32 - 25\log(\theta)dBi, & 1° \leq \theta \leq 48°, \\ -10\,dBi, & 48° \leq \theta \leq 180°, \end{cases} \tag{9.5}$$

where θ is the angle, in degree, from the axis of the main lobe, according to Figure 9.4.

9.5.3.2 WLL Antenna Patterns

The simulation of interference into FS-ML due to the WLL assumes a cell arrangement with 3 sectors. The base station antenna's one sector horizontal and vertical plane pattern is given by Sinclair Techn. Ltd [21]

$$G_{\mathrm{BS,Hor}}(\theta) = \begin{cases} 15 - 10\log\left[\exp(2.018\theta/120°)^{2.972}\right]\mathrm{dBi}, & 0° \leq \theta \leq 120° \\ -20\,\mathrm{dBi}, & 120° \leq \theta \leq 180° \end{cases} \qquad (9.6)$$

$$G_{\mathrm{BS,Ver}}(\theta) = \begin{cases} 15 - 10\log\left[\exp(4.58\theta/30°)^{1.37}\right]\mathrm{dBi}, & 0° \leq \theta \leq 30° \\ -20\,\mathrm{dBi}, & 30° \leq \theta \leq 180° \end{cases} \qquad (9.7)$$

Fixed users in the WLL system may use two different antennas; an omni-directional antenna with gain $G_U = 0\,\mathrm{dBi}$, or a directed antenna, whose radiation pattern is given in Figure 9.5 [3].

9.5.4 Traffic and Capacity Analysis

The number of sites, or base stations, required in the region over the WLL planning can be expressed as

$$N_{\mathrm{BS}} = \max\left\{\left\lceil\frac{A_0}{R_0^2\pi}\right\rceil, \left\lceil\frac{E_{\mathrm{Tot}}}{E_{\mathrm{Sec}}N_{\mathrm{Sec}}}\right\rceil\right\} \qquad (9.8)$$

where the maximization is over the coverage and capacity constraints, under assumption that there is one carrier only. The A_0 is the region area, R_0 is the cell radius, according to the link budget, E_{Tot} is the maximum total *Erlang* requirement for the region, and E_{Sec} is the maximum *Erlangs* per sector in the single cell with number of sector N_{Sec}.

Having in mind that in the WLL there is no roaming and handoff process, the number of traffic channels per sector is given by

$$K_{\mathrm{Tr,Sec}} = F_B(E_{\mathrm{Sec}}, \mathrm{GOS}) \qquad (9.9)$$

where $F_B(.)$ denotes *Erlang-B* function [23], which returns the number of channels given the *Erlang* requirements, and of grade of service (GOS).

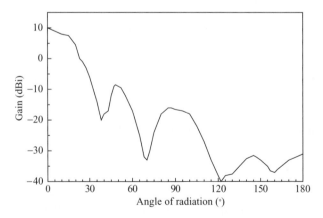

Figure 9.5 Antenna patttern for WLL subscriber unit

9.5.5 The WLL Channel Attenuation and Coverage Margin

The basic criterion used for defining system reliability and coverage margin is that a given signal level has to be exceeded in $Q\%$ of the cell area. In our model, the signal propagation inside the WLL cell is affected by the propagation attenuation and shadowing of the WLL channel, disregarding fast fading.

For a particular WLL base station—subscriber unit pair the channel attenuation is the same in the reverse link as it is in forward link. The channel attenuation is assumed to have a propagation attenuation exponent ξ, and is subject to log-normal shadowing.

The mean signal level is a stochastic variable varying slowly in time, with log-normal probability density function

$$pdf(\mu/r) = \frac{1}{\sqrt{2\pi}\sigma} \exp\left\{[\mu - \mu(r)]^2/2\sigma^2\right\} \tag{9.10}$$

where μ is the predicted mean signal level, r is the distance from the base station and σ is standard deviation whose typical value is 8 dB, [3].

If we assume that the propagation attenuation is increasing with the ξ-th power of distance, the mean received power in dB, can be expressed as

$$\mu(r) = \mu(R_C) + 10\xi \log(R_C/r) \tag{9.11}$$

where R_C is the cell radius. Further, if we assume the uniform distribution of users throughout the cell, the overall probability that the receiver minimum sensitivity, μ_{min}, is exceeded is given by

$$P_{ROB,Sens}(m_S) = \int_0^{R_C} \int_{\mu_{min}}^{\infty} \frac{2r}{R_C^2} p\, df(\mu/r)\, d\mu dr = \int_0^{R_C} \frac{r}{R_C^2} \text{erfc}\left[\frac{m_S + 10\xi \log(r/R_C)}{\sqrt{2\pi}\sigma}\right] dr \tag{9.12}$$

where $m_S = \mu(R_C) - \mu_{min}$ represents shadowing, or coverage, margin.

9.5.6 Power Control

Due to an interference limited capacity of a CDMA WLL system, an accurate power control must be active, which means that all subscriber unit signals must arrive at the same power at the base station.

In our WLL system model we have used the simple power control algorithm described by W. C. Y. Lee [15], meeting the requirement that the base station signal can still reach the subscriber unit at distance r, from the cell site with a reduced power.

We assumed that the power control includes both of the FWL and RVL. The power control laws are:

$$P_{FWL}(r) = \begin{cases} 0.55P_p, & 0 < r \le 0.55R_0 \\ \left(\dfrac{r}{R_0}\right)^2 P_p, & 0.55R_0 < r \le R_0 \end{cases} \tag{9.13}$$

$$P_{RVL}(r) = \left(\frac{r}{R_0}\right)^2 P_M, \quad 0 < r \le R_0 \tag{9.14}$$

where P_P is the maximal cell site transmitter power, and P_M is the maximal power emitted by a user. Indices FWL and RVL stand for forward link and reverse link, respectively.

9.5.7 WLL Links Budget

The reverse link budget gives the estimate of the maximum path-loss between the subscriber unit and the base station, for which the required $E_b/(N_0 + N_I)$ can be achieved, where E_b is the received bit energy at the base station, while N_0 and N_I are the p.s.d. of AWGN and interference, respectively. The N_I includes the intracell interference only, having in mind practically a very rarely clasterization of the WLL system cells.

The calculation is performed for the average number of users in service per cell, K_{Tr}, according to the traffic offered. The minimum signal power at the base station receiver input per sector, can be derived as

$$P_{BS,Rx-min,Sec} = \frac{\left(\dfrac{E_b}{N_0 + N_I}\right) N_0 B_{SS}}{\dfrac{B_{SS}}{V_b} - \left(\dfrac{E_b}{N_0 + N_I}\right)\alpha(K_{Tr,Sec} - 1)} \tag{9.15}$$

where B_{SS} is the spreading bandwidth and α is the speech activity factor.

The task of the forward link budget is to estimate the necessary base station transmitter power. Assuming the uniform distribution of subscriber units inside the cell, and following the power control algorithm, the average base station transmitter power per sector, can be obtained as [15]

$$P_{BS,Tx-min,Sec} = \int_0^{R_C} \int_0^{2\pi/N_{Sec}} P_{R_C} \frac{\alpha K_{Tr,Sec}}{R_C^2 \pi} \frac{r^2}{R_C^2} r \, dr \, d\varphi = P_{R_C} \frac{\alpha K_{Tr,Sec}}{2N_{Sec}} \tag{9.16}$$

where P_{R_c} is the power required to reach the unit at the cell boundary R_C.

9.5.8 Signals Description

The total signal at the FS-ML receiver input is

$$u(t) = u_d(t) + u_I(t) + n(t) \tag{9.17}$$

where the first term, $u_d(t)$, represent the desired FS-DML or FS-FDM/FM signal given by

$$u_{DML}(t) = \sqrt{2P_0} \, d(t) \cos(\omega_0 t) \tag{9.18}$$

and

$$u_{FM}(t) = \sqrt{2P_0}\cos[\omega_0 t + \varphi(t)] \quad (9.19)$$

with the mean power P_0, carrier frequency $f_0 = \omega_0/2\pi$.

The signal $d(t)$ in Equation (9.18) is the shaped modulating digital signal according to the type of digital modulation. The symbol rate and duration of $d(t)$ are V_s and T_s, respectively.

The instantaneous phase deviation of FM signal is

$$\varphi(t) = 2\pi \int [x_{FDM}(\tau) \otimes h_p(\tau)]d\tau \quad (9.20)$$

where $x_{FDM}(t)$ is the modulating FDM signal and $h_P(t)$ is the preemphasis pulse response. Symbol $\otimes$ stands for convolution.

The second term in expression (9.17)

$$u_I(t) = \sum_k \sqrt{2P_{I,k}} b^{(k)}(t - \tau_k) c^{(k)}(t - \tau_k) \cos[\omega_I t + \theta_k] \quad (9.21)$$

represents the interfering SSDS-CDMA signal origins from WLL users or WLL base stations. Its carrier frequency offset is $f_\Omega = \Omega/2\pi = (\omega_I - \omega_0)/2\pi$, τ_k is the time delay uniformly distributed in the interval $[0, T_b]$, θ_k is the random phase of each WLL signal uniformly distributed over $[0, 2\pi)$.

The modulating signal $b^{(k)}(t)$ and pseudorandom sequence $c^{(k)}(t)$ are of the following form:

$$b^{(k)}(t) = \sum_i b_i^{(k)} \prod(t - iT_b), \quad b_i^{(k)} \in \{\pm 1\} \quad (9.22)$$

$$c^{(k)}(t) = \sum_i c_i^{(k)} \prod(t - iT_c), \quad c_i^{(k)} \in \{\pm 1\} \quad (9.23)$$

The signal $\prod(.)$ is the rectangular pulse of unit amplitude and of duration $T_b = 1/V_b$, or $T_c = 1/V_c$, where V_b and V_c are bit and chip rate, respectively.

The processing gain of SS-CDMA WLL system is $G_P = V_c/V_b$.

In expression (9.21) $P_{I,k}$ is the power of the kth WLL interference source at the FS-ML receiver input, and is given by

$$P_{I,k \to FS} = P_k G_k(\theta_{k \to FS}, \varphi_{k \to FS}) G_{FS}(\theta_{FS \to k}, \varphi_{FS \to k}) A(d_{k \to FS}). \quad (9.24)$$

In above expression,

- P_k is the power of the radiated signal from the terminal or the base station.
- $G_k(\theta_{k \to FS}, \varphi_{k \to FS})$ is the gain from the source antenna in direction of the antenna FS.
- $\theta_{k \to FS}, \varphi_{k \to FS}$ are the relative co-ordinates of the straight line in spherical co-ordinate system whose referent point of the co-ordinate system is on the location of antenna FS.

- $G_{FS}(\theta_{FS \to k},\ \varphi_{FS \to k})$ is the gain of the antenna FS in the direction of the source on interference.
- The loss A is the loss of the signal as a function of distance.

The third component in Equation (9.17) is the *additive white Gaussian noise* (AWGN). One-sided *power spectral density* (p.s.d.) of this noise is N_0.

It is assumed that all the signals at the receiver input are mutually statistically independent.

9.6 Spectral Characteristics of Signals

9.6.1 FDM/FM Signal Power Spectral Density

In calculating the interference noise at the FDM/FM receiver output, it is necessary to know the p.s.d. of the FM signal. The FDM/FM signal has been assumed to carry multichannel telephone signals, described by bandlimited Gaussian noise with p.s.d.

$$S_{FDM}(f) = \begin{cases} \dfrac{(\Delta f_{rms})^2 |H_P(jf)|^2}{2(f_2 - f_1)}, & f_1 \le |f| \le f_2, \\ 0, & \text{for other frequencies} \end{cases} \tag{9.25}$$

where f_1 and f_2 are the lowest and highest baseband frequencies, respectively, Δf_{rms} is the r.m.s. of the total frequency deviation and $H_P(jf)$ is the preemphasis transfer function given by CCIR [4]

$$|H_P(jf)|^2 = 0.4 + 0.8(f/f_2)^2 + 1.3172(f/f_2)^3, \quad f_1 \le f \le f_2 \tag{9.26}$$

The normalized *low frequency equivalent* (l.f.e.) of the p.s.d. of a FDM/FM signal may be written as

$$S_{FDM/FM}(f) = \exp\left[-2 \int_{f_1}^{f_2} \frac{S_{FDM}(f)}{f^2}\,df\right] \sum_{n=0}^{\infty} \frac{1}{n!}\left[\frac{S_{FDM}(f)}{f^2} \overset{n}{\otimes} \frac{S_{FDM}(f)}{f^2}\right] \tag{9.27}$$

where symbol $\overset{n}{\otimes}$ stands for convolution, and has the following meaning:

$$z(f)\overset{n}{\otimes}z(f) = \begin{cases} \delta(f), & n = 0, \\ z(f), & n = 1, \\ n\text{-th convolution} & \text{for } n > 1. \end{cases} \tag{9.28}$$

In expression (9.27), the term

$$2\int_{f_1}^{f_2} \frac{S_{FDM}(f)}{f^2}\,df \tag{9.29}$$

is the mean square phase deviation.

At last, taking into account the output filter from the FM transmitter, l.f.e. of p.s.d. of the signal at the output of the FDM/FM transmitter is given by expression

$$S_{FM}(f) = S_{FDM/FM}|H_{FM}(jf)|^2 \tag{9.30}$$

where $H_{FM}(jf)$ is the transfer function of the output filter.

The frequency bandwidth of FDM/FM signal is given by expression [4]

$$B_{FM} = 2(n_v \Delta f_{eff} + f_2) \tag{9.31}$$

where n_v is a peak factor.

Having in mind that the l.f.e. of the p.s.d. of an FDM/FM signal can be expressed as the sum of a residual carrier and a series of terms involving convolutions of the p.s.d. of modulating signal, the p.s.d. of the FDM/FM signals considered have been evaluated using a finite number of terms of the expansion (9.27).

Using a fast convolution procedure, based on the Fast Fourier Transform, $S_{FM}(f)$ has been computed according to [18]. Finite number N of convolution terms with a suitable number M of discrete samples of the baseband signal and the corresponding spacing f_d between adjacent samples have been chosen, so as to make the truncation error very small. For a 960 channel system $N = 16$, $M = 41$, $f_d = 100\,\text{kHz}$; for an 1800 channel system $N = 8$, $M = 83$ and $f_d = 100\,\text{kHz}$.

The plot of the results of computation of $[S_{FM}(f)f_2]$, in a decibel scale, for 960 and 1800 channel systems are shown in Figure 9.6. A spectral line at the carrier frequency is always present and its normalized power is given by Equation (9.29).

The FDM/FM signals whose spectral characteristics we are interested are listed in Table 9.1, together with their main parameters. For the sake of clarity, the power level of the residual carrier, for 960 and 1800 channel systems, is not shown in Figure 9.6, but is reported in Table 9.1.

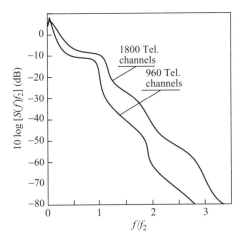

Figure 9.6 Normalized signal l.f.e. p.s.d.'s for the FM systems carrying 960 and 1800 telephone channels, versus normalized baseband frequency; f_2 is the highest baseband frequency. The output filter $H_{FM}(jf)$ is six poles Butterwoth's type, and minimum attenuation of 50 dB inout of band

Table 9.1 Main parameters of FDM/FM systems; Δf_0 and Δf_{eff} are test tone frequency deviation and r.m.s. of the total frequency deviation, respectively and m is a modulation index

Number of telephone channels	Baseband frequencies		Δf_0 (kHz)	Δf_{eff} (MHz)	m	B_{FM} (MHz)	Residual carrier power (dB)
	f_1 (kHz)	f_2 (kHz)					
960	60	4028	200	1.102	0.27	16.34	−9.36
1260	60	5636	200	1.262	0.22	20.76	−8.59
1800	316	8204	140	1.056	0.13	24.35	−0.89
2700	316	12388	140	1.293	0.10	34.50	−0.84

9.6.2 DML Signal Power Spectral Density

Under assumption that the normalized p.s.d. of the modulated signal at the input of the transmitter's filter is

$$S_{\text{DM}}(f) = \frac{1}{V_s}\left[\frac{\sin(\pi f/V_s)}{\pi f/V_s}\right]^2 \tag{9.32}$$

where V_S is the symbol rate.

The l.f.e. of p.s.d. of FS-DML signal at the filter output is given by

$$S_{\text{DML}}(f) = \begin{cases} \dfrac{P_0}{V_s}\left[\dfrac{\sin(\pi f/V_s)}{\pi f/V_s}\right]^2, & |f| \leq \dfrac{1-\rho}{2T_s} \\[3mm] \dfrac{P_0}{V_s}\left[\dfrac{\sin(\pi f/V_s)}{\pi f/V_s}\right]^2 \cos^2\dfrac{\pi[2f/V_s - 1 + \rho]}{2\rho}, & \dfrac{1-\rho}{2T_s} \leq |f| \leq \dfrac{1+\rho}{2T_s} \\[3mm] 0, & |f| > \dfrac{1+\rho}{2T_s} \end{cases} \tag{9.33}$$

where ρ is the roll-off factor and P_0 is the mean power of the modulated signal. The plot of $[S_{\text{DML}}(f)V_s/P_0]$ is shown in Figure 9.7 as a function of f/V_s.

9.6.3 SSDS-CDMA Signal Power Spectral Density

The generic form of SSDS-CDMA signal is given by Equation (9.21). The normalized low frequency equivalent (l.f.e.) of power spectral density of an SSDS signal (one user in WLL system) can be expressed as

$$S_{\text{SSDS}}(f) = [S_b(f) \otimes S_c(f)]|H_{\text{SSDS}}(jf)|^2 \tag{9.34}$$

where $H_{\text{SSDS}}(jf)$ is the transfer function of the transmitter output filter. In our analysis we suppose $H_{\text{SSDS}}(jf)$ is cosine filter, as in Equation (9.33).

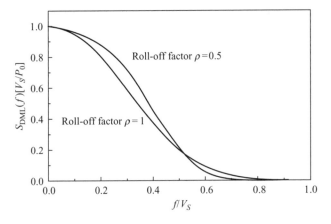

Figure 9.7 The normalized low frequency equivalent of power spectral density of FS/DML signal

The normalized power spectral density of binary random process $b(t)$ is

$$S_b(f) = \frac{1}{V_b} \left[\frac{\sin (\pi f/V_b)}{\pi f/V_b} \right]^2 \tag{9.35}$$

while the l.f.e. of p.s.d. of spreading sequence $c(t)$, with period L, is the following line spectrum:

$$
\begin{aligned}
S_c(f) = & \frac{\delta(f - f_\Omega)}{2L^2} + \frac{L+1}{2L^2} \sum_{\substack{k=-\infty \\ k\neq 0}}^{k=\infty} \left[\frac{\sin \pi(f - f_\Omega)/V_c}{\pi(f - f_\Omega)/V_c} \right]^2 \delta\left(f - f_\Omega - k\frac{V_c}{L} \right) \\
& + \frac{\delta(f + f_\Omega)}{2L^2} + \frac{L+1}{2L^2} \sum_{\substack{k=-\infty \\ k\neq 0}}^{k=\infty} \left[\frac{\sin \pi(f + f_\Omega)/V_c}{\pi(f + f_\Omega)/V_c} \right]^2 \delta\left(f + f_\Omega - k\frac{V_c}{L} \right)
\end{aligned}
\tag{9.36}
$$

However, for all practical purpose, when the period of a spreading sequence is $L \gg 1$ and $L \gg G_P$, the normalized l.f.e. of p.s.d. of an SSDS signal can be approximated as

$$
\begin{aligned}
S_{\text{SSDS}}(f) \cong & \frac{P_{\text{WLL-U}}}{V_b V_c} \int_{-\infty}^{\infty} \left[\frac{\sin(\pi\mu/V_b)}{\pi\mu/V_b} \right]^2 \left[\frac{\sin(\pi(f - f_\Omega - \mu)/V_c)}{\pi(f - f_\Omega - \mu)/V_c} \right]^2 |H_{\text{SSDS}}(jf)|^2 d\mu \\
& + \frac{P_{\text{WLL-U}}}{V_b V_c} \int_{-\infty}^{\infty} \left[\frac{\sin(\pi\mu/V_b)}{\pi\mu/V_b} \right]^2 \left[\frac{\sin(\pi(f + f_\Omega - \mu)/V_c)}{\pi(f + f_\Omega - \mu)/V_c} \right]^2 |H_{\text{SSDS}}(jf)|^2 d\mu
\end{aligned}
$$

$$\tag{9.37}$$

where $P_{\text{WLL-U}}$ is the mean power of the WLL user transmitter.

9.7 Interference Effects Analysis

9.7.1 Interference Noise Power Contribution

The total mean power at the FS-ML receiver input, contributed by all interference sources, taking into account their distribution, is given by the following relations:

$$P_{I,k} = P_{I,k \to FS} \Big|_{f \in B_I} S_{CDMA}(f) \, df \tag{9.38}$$

where B_I is the appropriate bandwidth at the FS-ML receiver input, and

$$P_{I,k \to FS} = \begin{cases} \sum_{j=1}^{N_B} \left(\iint\limits_{r,\varphi \in A_C} \frac{\alpha K_{Tr}}{A_C} P_{U,k} G_{U,k}\big[\theta_{U,k \to FS}(r,\varphi)\big] G_{FS}\big[\theta_{FS \to U,k}(r,\varphi)\big] \times CHA(r,\varphi) r \, dr d\varphi \right) \\ \text{WLL-users,} \\ \iint\limits_{r,\varphi \in A_0} \frac{N_B}{A_0} P_{BS,k} G_{BS,k}\big[\theta_{BS,k \to FS}(r,\varphi)\big] G_{FS}\big[\theta_{FS \to BS,k}(r,\varphi)\big] \times CHA(r,\varphi) r \, dr d\varphi \\ \text{WLL-Base stations.} \end{cases}$$

$$\tag{9.39}$$

In above expression,

- $P_{U,k}$ and $P_{BS,k}$ are the kth subscriber's unit and the kth base station transmitter power, respectively.
- The $G_{U,k}\big[\theta_{U,k \to FS}(r,\varphi)\big]$ and $G_{FS}\big[\theta_{FS \to U,k}(r,\varphi)\big]$ are each other antennas gain between the kth subscribers unit and FS-ML system with appropriate angle of radiation $\theta_{U,k \to FS}(r,\varphi)$ and $\theta_{FS \to U,k}(r,\varphi)$.
- $G_{BS,k}\big[\theta_{BS,k \to FS}(r,\varphi)\big]$ and $G_{FS}\big[\theta_{FS \to BS,k}(r,\varphi)\big]$ are each other antennas gain between the kth WLL base station and FS-ML system with appropriate angle of radiation $\theta_{BS,k \to FS}(r,\varphi)$ and $\theta_{FS \to BS,k}(r,\varphi)$ respectively.
- Value $CHA(r,\varphi)$ is a channel attenuation.
- (r,φ) are appropriate polar co-ordinates, according to a given co-ordinate system, Figure 9.4.

9.7.2 Interference Noise at the FDM/FM Receiver Output

The signal at the L-D input is given by expression [5]

$$u_M(t) = u_{FM}(t) + i_I(t) \otimes h_{IF}(t)$$

$$= Re\left\langle \sqrt{2P_0} \exp[j\omega_0 t + j\varphi(t)] \left\{ 1 + \sum_k I^{(k)}(t) \exp[-j\varphi(t) + j\theta_n] \right\} \right\rangle$$

$$= Re\left\langle \sqrt{2P_0} A(t) \exp[j\omega_0 t + j\varphi(t) + j\lambda(t)] \right\rangle \tag{9.40}$$

where $h_{IF}(t)$ is the IF filter pulse response, and

$$I^{(k)}(t) = \sqrt{\frac{P_{I,k}}{P_0}}[b^{(n)}(t - \tau_n)c^{(n)}(t - \tau_n)\exp(j\Omega t)] \otimes h_{IF}(t) \tag{9.41}$$

and,

$$\lambda(t) = \text{Im}\left\langle \ln\left\{1 + \sum_{k=1}^{K} I^{(k)}(t)\exp[-j\varphi(t) + j\theta_n]\right\}\right\rangle \tag{9.42}$$

At the L-D output we get the signal

$$u_D(t) = \frac{1}{2\pi}\frac{d}{dt}[\varphi(t) + \lambda(t)] \tag{9.43}$$

where the first component represents the desired signal and the second one is the interference noise.

Under assumption that in real working conditions the following is valid:

$$\left|\sum_{n=1}^{N} I^{(n)}(t)\right|_{\max} < 1 \tag{9.44}$$

the expression for the interference noise at the L-D output could be written in the following form:

$$\lambda(t) = \text{Im}\left\langle \sum_{m=1}^{\infty} \frac{(-1)^{m-1}}{m}\left\{\sum_{k=1}^{k} I^{(k)}(t)\exp[-j\varphi(t) + j\theta_n]\right\}^m\right\rangle \tag{9.45}$$

By applying the multinomial theorem [1], for the autocorrelation function of the interference noise $R_\lambda(\mu) = E\langle\lambda(t)\lambda^*(t + \mu)\rangle$, we find

$$R_\lambda(\mu) = \sum_{m=1}^{\infty} \frac{1}{4m^2} \sum_{m_1+\ldots+m_n=N} \left(\frac{m!}{m_1!\ldots m_N!}\right)^2$$
$$\times \left[\prod_{n=1}^{m_n} R_1^{m_n}(\mu)R_0^{m_n}(\mu)^* + \prod_{n=1}^{m_n} R_I^{m_n}(\mu)^* R_0^{m_n}(\mu)\right] \tag{9.46}$$

where

$$R_0^{(m_n)}(\mu) = E\langle\exp[jm_n\varphi(t) - jm_n\varphi(t + \mu)]\rangle \tag{9.47}$$

and

$$R_I^{(m_n)}(\mu) = E\left\langle\left[I^{(n)}(t)I^{(n)}(t + \mu)^*\right]^{m_n}\right\rangle \tag{9.48}$$

Taking into account relation (9.44), for the autocorrelation function of the interference noise, $R_\lambda(\mu) = E\langle\lambda(t)\lambda^*(t+\mu)\rangle$, we find

$$R_\lambda(\mu) \cong \frac{1}{8} Re \left\{ \sum_{n=1}^N E \left\langle R_I^{(m_n=1)}(\mu) R_0^{(m_n=1)}(\mu)^* \right\rangle \right\} \tag{9.49}$$

By applying the Wiener–Khintchine theorem to the autocorrelation of $\lambda(t)$, and taking into account relation (9.44), the interference noise p.s.d. at the FDM-FM receiver output is

$$S_{\mathrm{IN}}(f) \cong \frac{f^2}{4|H_P(jf)|^2} \frac{P_{I,k}}{P_0} \left\{ \left[S_{\mathrm{CDMA}}(f - f_\Omega)|H_{\mathrm{IF}}(jf)|^2 \otimes S_{\mathrm{FM}}(-f) \right] \right.$$
$$\left. + \left[S_{\mathrm{CDMA}}(-f - f_\Omega)|H_{\mathrm{IF}}(if)|^2 \otimes S_{\mathrm{FM}}(f) \right] \right\} \tag{9.50}$$

where $S_{\mathrm{CDMA}}(f)$ is the Fourier transform of the autocorrelation of interfering CDMA signal.

The interference noise power in a telephony channel centred at f_{ch} may be written as

$$N_I = \frac{2b S_{\mathrm{IN}}(f_{\mathrm{ch}})}{(\Delta f_0)^2} \,(\mathrm{mWp}) \tag{9.51}$$

where $b = 1.7\,\mathrm{kHz}$ is the telephone channel psophometric band and Δf_0 is the FM signal test tone deviation.

9.7.3 Probability of Error at the DML Receiver Output

For evaluating the interference effects on FS-DML due to the WLL, we used the formulas for the probability of error in AWGN, because the CDMA interference was considered as white noise. Generally, this is a questionable approximation, but the relatively flat p.s.d. of CDMA signal in the bandwidth of interest gives us a certain degree of confidence in the interference analysis.

If we denote the p.s.d. of total CDMA signal by

$$N_I = \frac{P_{I,k}}{B_d} \int_{f_\Omega - B_d/2}^{f_\Omega + B_d/2} S_{\mathrm{CDMA}}(f)\,\mathrm{d}f \tag{9.52}$$

where B_d is the bandwidth of DML signal, *BER* for M-QAM, at the output of DML receiver are given by

$$P_{e,M-\mathrm{QAM}} \cong \frac{1}{ld(M)} \left\{ 1 - \left[1 - \left(1 - \frac{1}{\sqrt{M}}\right) \frac{1}{2}\mathrm{erfc}\sqrt{\frac{3}{2(M-1)} \cdot \frac{E_{b,\mathrm{DMI}}ld(M)}{N_0 + N_I}} \right]^2 \right\} \tag{9.53}$$

where $E_{b,\mathrm{DML}}$ is the mean energy per bit of DML signal.

9.8 Study Case

The allocation of radio spectrum resources is coordinated at an international level through the ITU, which has established a strict set of rules referred to as the Radio Regulations, designating the types of services that may occur within each specific spectral region. To improve the efficiency of the frequency allocation procedure, including the electromagnetic coexsistence problems, frequencies may be reallocated and reused within other geographical regions providing each of the regions that share common frequencies.

The main parameters typical numerical values, in systems under consideration, used in computer simulation of interference into FS-ML due to the WLL system overlay, are given in Tables 9.2 and 9.3 [5].

To quantify the interference effects on the performance of the FS-ML we have used the criterion described in [7, 8]. For the FS-FDM/FM system the interference produced by SSDS-WLL has such a value that it causes no more than 1 dB degradation of the output *S/N* ratio, or the total interference power must be about 6 dB below the thermal noise level.

Table 9.2 FS-ML system parameters

Parameter type	Value
Frequency band	1.8/1.9 GHz
Antenna height	$H_{FS} = 100\,m$
Antenna gain	32 dBi
FDM telephone channels number	$N_{FDM} = 600$
Total r.m.s. frequency deviation	$\Delta f_{rms} = 0.871\,MHz$
Test tone deviation	$\Delta f_0 = 200\,kHz$
FDM bandwidth	$f_1 = 60\,kHz, f_2 = 2.54\,MHz$
Desired signal average power at the FM receiver input	$P_0 = 70\,nW$
IF filter in FM receiver	Butterworth, six poles
Power spectral density of AWGN	$F_{eq}kT = 10^{-19}\,W/Hz$
Interference noise power allowed in FS-FDM/FM	$-95\,dBm$
Equivalent bit rate in FS-DML 64QAM system	$V_b = 32\,Mb/s$
Rolloff factor	$\rho = 0.5$
BER in FS-DML 64QAM	10^{-6}
BER allowed in FS-DML 64QAM	10^{-5}

Table 9.3 Initial assumptions for the WLL system parameters

Parameter type	Value
Frequency band	1.8/1.9 GHz
WLL region radius	$R_0 = 30\,\text{km}$
Cell 'A' radius	$R_A = 5\,\text{km}$
Cell 'B' radius	$R_B = 2\,\text{km}$
Number of cells 'B' according to the region population	$N_B = 20$
Number of users in cell 'A'	$K_A = 1000$
Number of users in cell 'B'	$K_B = 300$
Traffic offered per user	0.15 Erlang
Grade of service	0.01
Voice activity factor	$\alpha = 0.5$
Propagation attenuation exponent	$\xi = 3.2$
Base station antenna gain	15 dBi
Base station antenna height	15 m
Fixed subscriber unit antenna gain	10 dBi
Average fixed subscriber unit antenna height	3 m
Bit rate	$V_b = 32\,\text{kb/s}$
Chip rate	$V_c = 10\,\text{Mchip/s}$
Receiver sensitivity	$\mu_{\min} = -117\,\text{dBm}$
Shadow margin	10 dB
Probability that $\mu_{\min}$ is exceeded	90 %
$E_b/(N_0 + N_I)$ required	4.5 dB

If we assume that in a typical suburban or rural area, near the high populated urban and industrial communities, the equivalent microwave noise factor is equal to F_{eq}, the total interference power level must be no greater than

$$I_{\text{FDM/FM}} = 10 \log\left(\frac{1}{4}\frac{F_{\text{eq}}kT_0 B_{\text{FM}}}{1mW}\right) \text{dBm} \qquad (9.54)$$

where $kT_0 = 4.10^{-21}$ W/Hz, and B_{FM} is the FM signal bandwidth.

For evaluating the performance degradation in FS-DML we have used relations (9.52–53). The required *BER* without interference influence is 10^{-6}. The maximum allowed *BER* due to SSDS-CDMA interference is 10^{-5}.

Computer simulation included interference arisen from WLL,

- Fixed subscriber units
- Base station.

The results obtained are shown in Figures 9.8–9.15, in a function of carrier frequency offset. In accordance to the system geometry, shown on Figure 9.4, the co-ordinate origin lies in the site of cell '*A*'. The location of FS-ML antenna is moved everywhere inside a

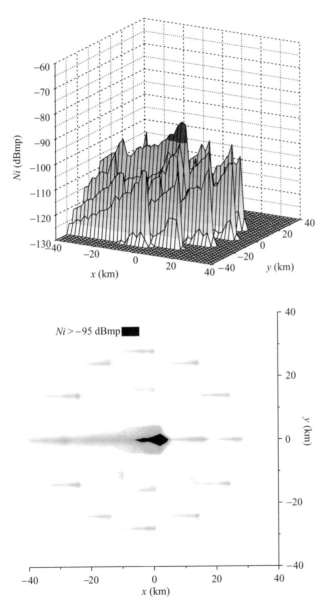

Figure 9.8 WLL subscriber units interference on FS-FDM/FM; $f_\Omega = 0$

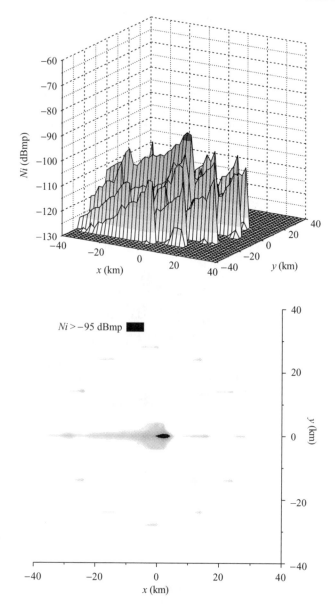

Figure 9.9 WLL subscriber units interference on FS-FDM/FM; $f_\Omega = 7.2$ MHz

square of $80 \times 80 \, \text{km}^2$ around the co-ordinate origin, while the antenna main beam is pointed always to the east.

In order to generalize the results, shown in Figures 9.8–9.15, many other simulation results are obtained using different sets of parameters, as different base antenna height, different systems regarding the bandwidth, user traffic and many others. Because of the quantity of graphic presented results, it is suitable to present them as the percentage of the total area where FS-ML can be located, regarding the interference level, as shown in Tables 9.4 and 9.5.

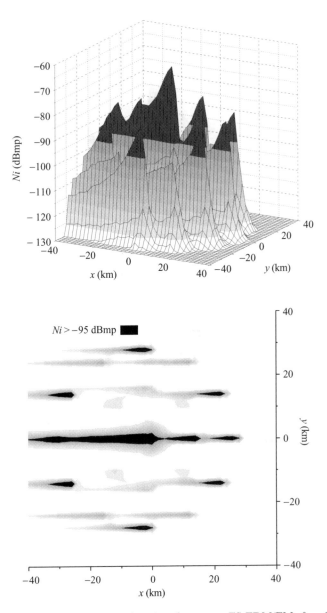

Figure 9.10 WLL base stations interference on FS-FDM/FM; $f_\Omega = 0$

The unallowed areas, are shown to be very small, practically negligible in all observed cases, in comparison to the total WLL region. It should be mentioned that only a small part of results are presented, illustrating typical sets of parameters with regard to worst case like chosen system parameters.

Considering all above-mentioned results the following conclusions are valid:

- The results obtained show that in the typical working conditions, the WLL overlay would not cause excessive interference into FS-ML, even when WLL operates in the same frequency band.

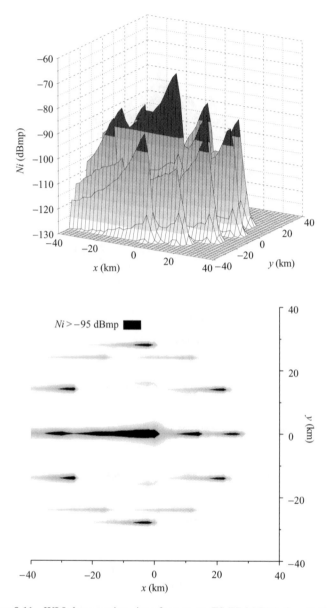

Figure 9.11   WLL base stations interference on FS-FDM/FM; $f_\Omega = 7.2 \, \text{MHz}$

- The interference in FS-ML system due to WLL reverse link overlay is considerably smaller comparing to the WLL forward link interference, and practically is neglected.
- FS-FDM/FM is more vulnerable on the WLL interference compared to FS-DML system
- By careful planning of the WLL base stations locations it is possible to realize full coexistence between the existing FS-ML and new overlaid WLL systems, without demanding new frequency bands.

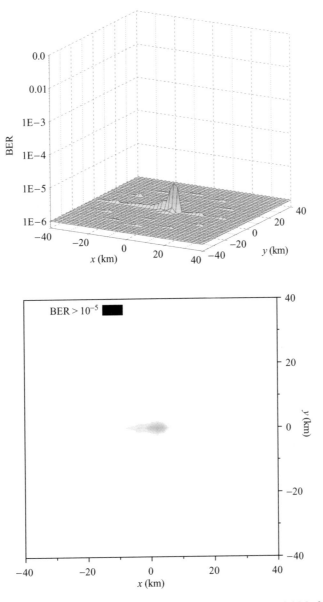

Figure 9.12 WLL subscriber units interference on FS-DML/64QAM; $f_\Omega = 0$

In general, when subscriber densities are low and the distances to the serving exchange are great, WLL systems based on point-to-point or point-to-multipoint radio technology currently offer the most appropriate technical solution. When servicing higher subscriber densities or extending the wireline network beyond its existing limits, WLL systems are more likely to be based on cellular and PCS radio technology, either analogue or digital.

The potential market is large, especially in the industrialized nations where the demand for high quality, rapidly deployable telecom service is the most acute, and in undeveloped regions where the necessity for telecommunication systems and services is so high.

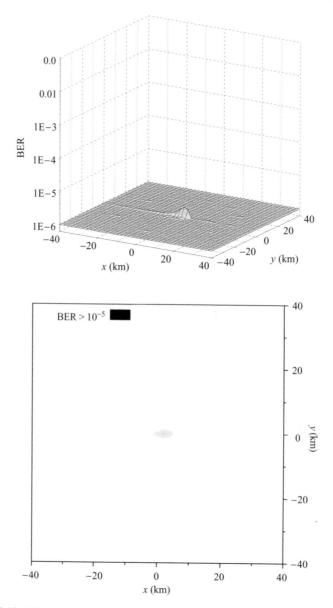

Figure 9.13   WLL subscriber units interference on FS-DML/64QAM; $f_\Omega = 7.2\,\text{MHz}$

SSDS-CDMA systems have been used by the military for a number of years due to their ability to resist interference and the practically neglected influence on other communication systems sharing the same frequency bands. In 1991 Qualcomm Inc. of San Diego, USA, proposed using CDMA for personal communication systems. The success of this system in a few previous years affected strong influence to the development and applications of this technology worldwide.

At this moment the SSDS-CDMA technology is practically seen the best solution for highly increasing demand for communications. Namely, the use of CDMA WLL technology

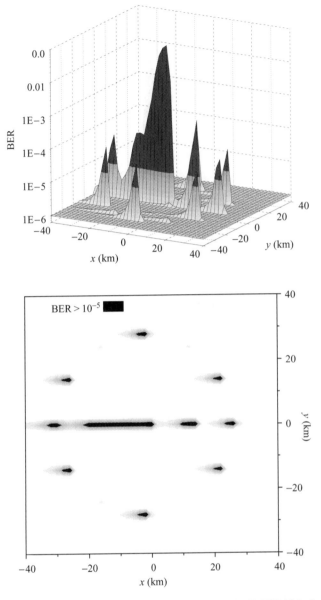

Figure 9.14 WLL base stations interference on FS-DML/64QAM; $f_\Omega = 0$

to provide telephone network access is dramatically growing. Its continued acceptance is being fueled by such key advances in technology as:

- Digital wireless technology offering improved voice quality, capacity, and security.
- Intelligent antenna technology, allowing extended coverage areas and reduced interference levels.
- Improved chip design and packaging for subscriber units, resulting in increased longevity and reduced per-subscriber costs.

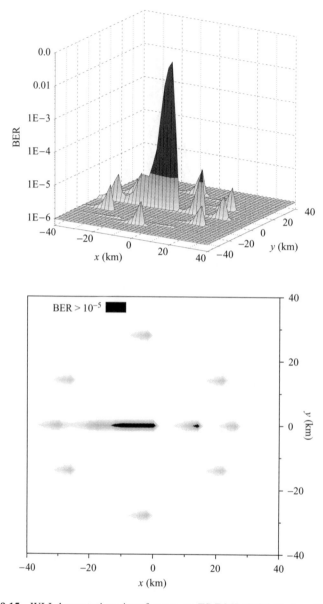

Figure 9.15 WLL base stations interference on FS-DML/64QAM; $f_\Omega = 7.2\,\text{MHz}$

- Increased integration of software-driven digital signal processing capabilities, allowing network recognition of multiple air-interface standards for improved compatibility and interoperability.
- Improved designs, allowing interference avoidance, spectrum sharing, and full scale-ability to promote use in both dense urban and sparse rural areas.

According to the above-mentioned facts, in this chapter the results of the interference analysis in FS-FDM/FM and FS-DML due to a typical example of new WLL systems

Table 9.4 Areas in WLL region with unallowed interference; FS-FDM/FM microwave links possible locations lie inside the area of $80 \times 80\,\mathrm{km}^2$, WLL region area is $2826\,\mathrm{km}^2$, while the WLL cells area is equal to $330\,\mathrm{km}^2$

Interference caused by	FS-ML antenna height is $H_{\mathrm{FS}}(\mathrm{m})$	FS-FDM/FM area, in %, with unallowed interference, respects to FS-FDM/FM region area — FDM = 600 channels				FS-FDM/FM area, in %, with unallowed interference, respects to FS-FDM/FM region area — FDM = 1260 channels			
		$V_c = 5.82\,\mathrm{Mch/s}$		$V_c = 12.46\,\mathrm{Mch/s}$		$V_c = 5.82\,\mathrm{Mch/s}$		$V_c = 12.46\,\mathrm{Mch/s}$	
		$f_\Omega = 0$	$f_\Omega = 4\,\mathrm{MHz}$	$f_\Omega = 0$	$f_\Omega = 7\,\mathrm{MHz}$	$f_\Omega = 0$	$f_\Omega = 7.2\,\mathrm{MHz}$	$f_\Omega = 0$	$f_\Omega = 10\,\mathrm{MH}$
WLL Reverse link [%]	50	2.4	1.2	0.8	0.4	2.7	2.1	1.8	0.4
	100	1.1	0.7	0.4	0.2	1.4	1.0	0.7	0.2
WLL Forward link [%]	50	5.1	4.2	4.3	3.0	5.5	4.9	5.2	3.0
	100	4.6	3.6	3.5	2.6	4.9	4.4	4.5	2.6

Table 9.5 Areas in WLL region with unallowed interference; FS-DML microwave links possible locations lie inside the area of $80 \times 80\,\mathrm{km}^2$, WLL region area is $2826\,\mathrm{km}^2$, while the WLL cells area is equal to $330\,\mathrm{km}^2$

Interference caused by	FS-ML antenna height is $H_{\mathrm{FS}}(\mathrm{m})$	FS-DML area, in %, with unallowed interference, respects to FS-DML region area. — DM = 512 channels				FS-DML area, in %, with unallowed interference, respects to FS-DML region area. — DM = 512 channels			
		$V_c = 5.82\,\mathrm{Mch/s}$		$V_c = 12.46\,\mathrm{Mch/s}$		$V_c = 5.82\,\mathrm{Mch/s}$		$V_c = 12.46\,\mathrm{Mch/s}$	
		$f_\Omega = 0$	$f_\Omega = 4\,\mathrm{MHz}$	$f_\Omega = 0$	$f_\Omega = 7\,\mathrm{MHz}$	$f_\Omega = 0$	$f_\Omega = 7.2\,\mathrm{MHz}$	$f_\Omega = 0$	$f_\Omega = 10\,\mathrm{MH}$
WLL Reverse link [%]	50	0.3	0.2	0.1	0.1	0.2	0.2	0.1	0.1
	100	0.1	0.1	0.0	0.0	0.1	0.0	0.0	0.0
WLL Forward link [%]	50	2.5	2.2	2.2	1.4	2.1	2.0	2.1	1.5
	100	2.1	1.7	1.8	0.5	1.2	1.1	1.5	0.5

with wide-band SSDS-CDMA techniques, sharing the same frequency band, is evaluated. The results obtained show that in the typical working conditions with careful cell planning, interference can be totally neglected, and in most situations WLL overlay would not cause excessive interference in the FS-ML.

This conclusion is of great importance for the development of telecommunication systems in suburban and rural areas especially, because it is possible to develop telecommunication systems fast and economically, without any additional frequency demands and under favourable financial conditions.

The results obtained are very important when we have in mind prospective progress in PCS. From the analysis presented in literature, one can see that the future PCS and WLL will certainly use higher frequency bands, in the first place 3 and 5 GHz, already occupied by numerous FS-ML. The possibility of overlay of new PCS and WLL based on wideband SSDS-CDMA technology, not introducing excessive interference in a FS-ML is very attractive, because it allows rapid, reliable and inexpensive solution of fast growing requirements in communications.

References

[1] A. Abramovitz and I. Stegun, *Handbook of Mathematical Functions*, Dover, New York, 1972.
[2] F. Adachi, M. Sawahashi and H. Suda, 'Wideband DS-CDMA for Next-Generation Mobile Communications Systems,' *IEEE Commun. Mag.*, vol. 36, no. 9, pp. 56–69, 1998.
[3] Q. Bi and D. R. Pulley, 'The Performance of DS-CDMA for Wireless Local Loop,' in *IEEE 4th International Symposium on Spread Spectrum Techniques and Applications*, Mainz, Germany, pp. 1330–1333, 1996.
[4] CCIR (1982), vol. IX, Geneve.
[5] M. L. Dukic and M. Babovic, 'Interference Analysis in Fixed Service Microwave Links due to Overlay of Broadband SSDS-CDMA Wireless Loop Systems,' *Wireless Network*, no. 6, pp. 109–119, J. C. Baltzer AG, Science Publisher, 2000.
[6] M. L. Dukic and Z. S. Dobrosavljevic, 'Interference Noise Evaluation in FDM/FM Radio System due to SS-CDMA PCS,' in *IEEE International Symposium on Spread Spectrum Techniques and Applications*, Mainz, Germany, 1996.
[7] EIA 10E/TIA Telecomm. Sys. Bull. TSB10E Interference criterion of microwave systems in the private radio services, 1990.
[8] EIA/TIA/IS-95 Mobile Station Compatibility Standard for Dual Mode Wideband Spread Spectrum Cellular Systems, 1993.
[9] C. C. Ferris, 'Spectral Characteristics of FDM-FM Signals,' *IEEE Trans. Commun.*, vol. COM-16, no. 2, pp. 233–238, 1968.
[10] K. G. Filis and S. C. Gupta, 'Overlay of Cellular CDMA on FSM,' *IEEE Trans. Vehicular Technology*, vol. 43, no. 1, pp. 86–98, 1994.
[11] A. Fukasawa, T. Sato, Y. Takizawa, T. Kato, M. Kawabe and R. Fisher, 'W-CDMA System for Personal Radio Communications,' *IEEE Commun. Mag.*, vol. 34, no. 10, pp. 116–123, 1996.
[12] H. Holma and A. Toskala *WCDMA for UMTS—Radio Access For Third Generation Mobile Access Communications*, John Wiley & Sons, New York, 2000.
[13] InterDigital Communication Corporation, *TrueLink System Description*, Mesto, USA, 1996.
[14] R. Kohno, R. Meidan and L. Milstein, 'Spread Spectrum Access Methods for Wireless Communications,' *IEEE Commun. Mag.*, vol. 33, no. 1, pp. 58–67, 1995.
[15] W. C. Y. Lee, 'Overview of Cellular CDMA,' *IEEE Trans. Vehicular Technology*, vol. 40, pp. 291–302, 1991.

[16] T. Magedanz, 'Integration and Evaluation of Existing mobile Telecommunications Systems toward UMTS,' *IEEE Commun. Mag.*, vol. 34, no. 9, pp. 90–97, 1996.

[17] P. J. Marshall, K. W. Sowerby and M. Shafi, 'The Feasibility of Spectrum Sharing Between DS-CDMA Mobile Radio Systems And Microwave Point-To-Point Links,' *IEEE Vehicular Technology Conference, VTC '96*, Atlanta, GA, USA, 1996.

[18] L. Moreno, 'Spectral Characteristics and Channel Spacing Criteria in FDM/FM Radio Links,' CSELT Rapporti Tecnici, vol. V, no. 3, pp. 141–148, 1997.

[19] SAMSUNG Electronics Co., LTD *Introduction of B-CDMA Wireless Access System*, Seoul, Korea, 1997.

[20] D. L. Schilling, R. L. Pickholtz, V. Erceg, M. Kulback, E. G. Kanterakis, W. H. Biederman and D. C. Salerno 'On the Feasibility of a CDMA Overlay for Personal Communications Networks,' *IEEE J. Selected Areas Commun.*, 1992.

[21] Sinclair Techn. Ltd, *Antenna Cataloge*, Cambridge, UK, 1997.

[22] Y. R. Tsai and J. F. Chang 'Feasibility of Adding a Personal Communications Network to an Existing Fixed Services Microwave System,' *IEEE Trans. Commun.*, vol. 44, no. 1, pp. 76–83, 1996.

[23] A. J. Viterbi, *CDMA Principles of Spread Spectrum Communications*, Addison Wesley, Massachusets, 1995.

Part II

Applications

10

Development of a Prototype of the Broadband Radio Access Integrated Network

Masugi Inoue, Gang Wu and Yoshihiro Hase

10.1 Introduction

Wireless local loop (WLL) or fixed wireless access systems are attractive solutions to the so-called last mile problem because of their low capital cost, fast network deployment capability and low maintenance cost [1]. They would, therefore, be effective during the transition period from *fibre to the curb* (FTTC) to *fibre to the home* (FTTH). WLL has the potential of becoming a major competitor of local exchange networks, *cable television* (CATV) networks, *digital subscriber line* family (xDSL), and FTTH. Actually, the *federal communications commission* (FCC) in the U.S.A. licensed 1 GHz of spectrum at 28 GHz and an additional 300 MHz of spectrum at 31 GHz for *local multipoint distribution service* (LMDS) systems. LMDS systems use mm-wave signals in those bands to transmit voice, video, and data signals within cells 3–10 miles in diameter. In Europe, the *European Telecommunications Standards Institute* (ETSI) has started a project for standardization of LMDS-like system, called HIPERACCESS, for physical and *data link* control (DLC) layers. In Japan, the Ministry of Posts and Telecommunications has announced that in the first stages of spectrum allocation, in total, 2040 MHz of spectrum at 22, 26 and 38 GHz bands will be allocated to service providers offering point-to-point and point-to-multipoint wireless access services. In addition to these *broadband* systems, attention is being paid to *narrowband* WLL systems that are based on the Japanese digital cordless phone system, called *personal handy-phone system* (PHS). WLL technology is also gaining popularity in the Asian and Latin American countries as a means of providing telephone services in sparsely populated rural areas.

The millimetre (mm) wave has become attractive for communications due to its potential for high-capacity transmission. The *Communications Research Laboratory* (CRL) has recently proposed a system concept called the *broadband radio access integrated network* (BRAIN) in the mm-waveband [2], that operates in the mm-wavebands such as those around 38 or 60 GHz, which are still undeveloped. BRAIN systems can be used as indoor high-speed wireless LANs or as outdoor broadband wireless access systems that serve as the last hop of FTTC.

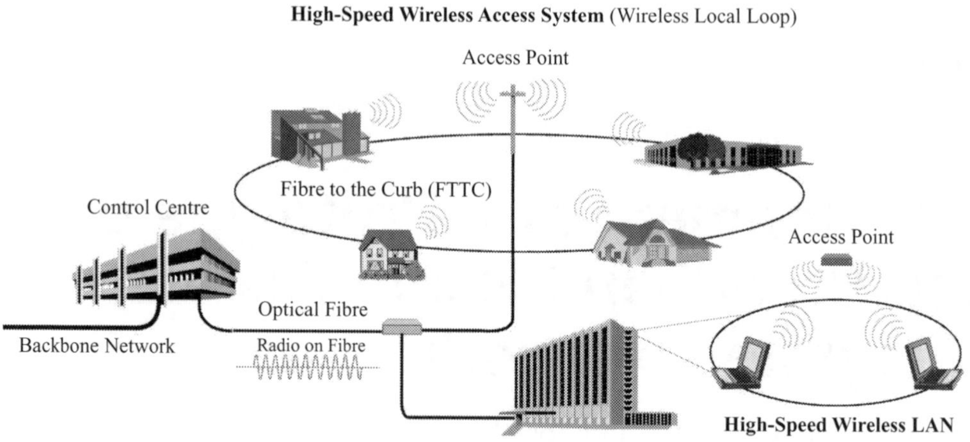

High-Speed Wireless Access System (Wireless Local Loop)

Figure 10.1   System configuration of BRAIN

The system configuration of BRAIN is shown in Figure 10.1. Indoor and outdoor *access points* (APs) are connected via optical fibre links to a control centre where signal processing and network switching is done. The APs only need to have an *optical/ electrical* (OE) converter because BRAIN incorporates optical fibre and mm-wave technologies such as mm-waveband signal generation that uses a fibre–optic frequency-tunable comb generator [3] and mm-waveband signal transmission on fibre (radio on fibre) [4]. These simple APs provide as easy and economical way to make broadband networks.

In this paper, we focus on the indoor system of BRAIN, a *mm-waveband high-speed multimedia wireless LAN*, and introduce its prototype designed and developed at CRL. Section 2 outlines the architecture and design of the prototype. Section 3 is a description of a wireless MAC protocol, and Section 4 describes the configuration and implementation of the system.

10.2   System Overview

10.2.1   Architecture

The indoor system of BRAIN offers broadband radio access services in an indoor environment. The total service area of the system, e.g. a large office room, is divided into a number of *basic service areas* (BSA), each including an access point (AP) and a number of fixed and/or quasi-fixed *stations* (ST). The radius of the BSA can be from 7 or 8 m to 80 or 90 m, depending on factors such as the link budget design for mm-wave communications and whether it is a furnished or unfurnished environment [5]. Each ST is connected to a multimedia (voice, data, and video) terminal and communicates with other ST(s) inside and/or outside of the BSA. Because an ST usually employs a directional antenna in mm-waveband radio communications, it should communicate with others via the AP. Traffic generated from or arriving at the BSA passes through the AP, and thus, the indoor system of BRAIN is a centralized control system.

Transmissions between AP and STs can be either on the same frequency band using *time division duplex* (TDD) or on separate frequency bands using *frequency division duplex* (FDD). TDD has many advantages, such as being able to support asymmetric traffic. However, it is still difficult to use it in a burst modem that has one-way transmission speed of 100 Mbps or higher. Therefore, we chose FDD for high-speed transmission in the current phase. The downlink channel is a *time division multiplexing* (TDM)-like channel on which the bit stream always flows, while the uplink channel is a *time division multiple access* (TDMA)-like channel on which the bit stream is sent in bursts from different STs.

10.2.2 Design Issues

In general, there are two kinds of wired network services: *transmission control protocol/internet protocol* (TCP/IP)-based Internet services and asynchronous transfer mode (ATM)-based services. It is, therefore, desirable to develop a wireless technology, that is applicable to these two types of wired networks. The additional functions that are needed are shown in Figure 10.2.

There are two major wireless multimedia network research topics: wireless access and mobility management [6–9]. The mobility management issues will not be considered here because mobile communication in the mm-waveband is still difficult to support. The wireless access issues include physical, *media access control* (MAC) and *data link control* (DLC) layers and wireless control functions.

The physical layer equipment should support burst transmission as well as provide high-speed and high-quality transmission. In order to make the expensive and undeveloped mm-wave communications feasible, it is important at this point to design simple and

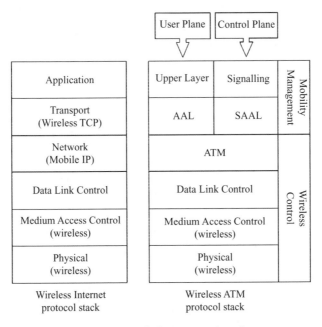

Figure 10.2 Wireless protocol stacks

economical physical-layer equipment. For instance, we can use a very simple modem with direct carrier modulation such as such as *On-Off Keying* (OOK) or *Frequency Shift Keying* (FSK).

10.3 Mac Protocol: RS-ISMA

10.3.1 Overview

Here, we introduce a wireless multimedia communications MAC protocol called *slotted idle signal multiple access* with reservation (RS-ISMA) [10] on which the functions of retransmission and wireless control can be easily implemented. RS-ISMA is a combination of reserved ISMA (R-ISMA) [11] and slotted ISMA (S-ISMA) [12]. Previous research on R- and S-ISMA has shown that these protocols have high throughput without any hidden terminal problems and can support integrated transmission. RS-ISMA consists of two steps: reservation and information transmission. We will describe the frame structure, slot configuration and the control signals used in the protocol and then outline the reservation and information transmission procedures.

10.3.2 Frame Structure

Figure 10.3 shows the structure of a MAC frame. It consists of a header and a body. The header includes frame control, addresses of source ST, destination ST and AP, a short message, and a *cyclic redundancy check* (CRC) code. A number of higher-layer *protocol data units* (PDU) or ATM cells and CRC code are included in the frame body. MAC frames are classified into two types: *control and management frames* (CMF) and *data frames* (DF). CMFs only have frame headers. DFs that have a variable numbers of PDUs have frame headers and bodies. The frame control field of frame header indicates whether frame is a CMF or DF. Thus, a CMF is short and has a fixed length, and a DF has a variable length.

10.3.3 Slot Configuration

The downward channel bit stream is organized into time slots. As shown in Figure 10.4, a time slot consists of two fields: a *control signal field* (CSF) and a *downstream information*

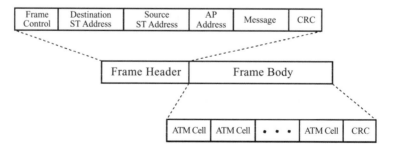

Figure 10.3 MAC frame structure

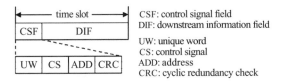

Figure 10.4 Time-slot configuration

field (DIF). The CSF at the beginning of a slot is used to broadcast control signals that control the traffic on the upward channel, while the DIF following the CSF is used to transmit downstream MAC frames. In general, CSF can be inserted between downstream MAC frames.

The CSF consists of four parts: a unique word, control signal, ST address, and CRC code. The control signal is the major part of the CSF. There are six control signals, which are defined later in details. The address of a specific ST, which is requested in order to transmit a DF or an ACK frame based on a polling scheme, is in the ST address part. The length of a time slot is such that a CMF can be sent during a time slot. The chosen length reflects a trade-off in efficiency between the upward and the downward channels [9].

10.3.4 Definition of Control Signals

An ST decides whether to start a transmission, or, whether to continue the current transmission according to a control signal inserted in a slot. The definition of each control signal is given as follows:

1. IDLE (idle) allows STs having a CMF to transmit based on a contention-based scheme.
2. POLL (polling) requests a specified ST to transmit a DF. The ST will transmit a DF, or a null frame (without frame body) when a DF is not ready.
3. ACKR (acknowledgement request) requests a specified ST to transmit an ACK frame acknowledging a downstream DF.
4. BUSY (busy) announces that the upward channel is busy.
5. CONT (continue) encourages an ST transmitting a DF in non-periodic polling mode to continue its transmission of another DF.
6. STOP (stop) forces an ST transmitting a DF to stop the transmission immediately.

10.3.5 Reservation Procedure

In RS-ISMA, an ST should transmit a *reservation packet* (RP) that belongs to CMF before beginning information transmission. The reservation procedure is based on the following contention-based scheme.

Figure 10.5 shows a time chart of the reservation procedure. The AP broadcasts an IDLE periodically when the upward channel is idle. In response to an IDLE, the STs that are to begin information transmission are allowed to contend for access to the channel by transmitting an RP. If an RP is received successfully, the AP sends an ACK in DIF within a given time-out period, and then, the information transmission procedure begins. If an ST that has sent an RP does not receive an ACK during the time-out period, the

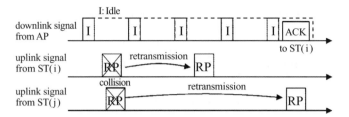

Figure 10.5   Reservation procedure

reservation fails. The transmission of an RP will fail if simultaneous transmission from different STs results in a collision, or if an RP is received with error. In both cases, the ST will retransmit an RP with some probability after hearing an IDLE again.

10.3.6   Information Transmission Procedure

After a successful reservation, an ST can transmit DFs based on a polling. A polling scheduler in the AP schedules the polling. It uses a polling table that contains the polling cycle, priority, the next scheduled polling time, and so on of each ST that succeeded in being reserved. This information is updated as the polling continues. *Periodic* and *non-periodic* polling modes support different traffic services.

10.3.6.1   Periodic Polling Mode

In periodic polling shown in Figure 10.6, the AP polls an ST by sending POLLs according to a polling cycle requested in reservation stage. After an ST receives a POLL addressed to itself, it immediately transmits a DF in which the number of PDUs included in the frame body is given in the frame header. In the case where a POLL is received but no DF is in the transmission buffer, the ST sends a null frame (without frame body) in order to keep the connection. If the frame header is checked out without any error, the AP inserts a BUSY in the CSF of the following slots, which announces that the upward channel is busy until the end of the current transmission. Otherwise, the AP inserts a STOP in the CSF, which forces the ST to stop the transmission. The AP updates the scheduler by calculating the next polling time according to the polling cycle after a successful transmission.

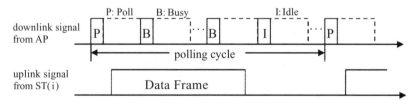

Figure 10.6   Periodic polling mode

10.3.6.2 Non-periodic Polling Mode

In the case of non-periodic polling shown in Figure 10.7, the transmission request involves a number of DFs, each with several PDUs, that are already in the transmission buffer. The AP polls STs with low priority between periodic high-priority pollings. When a POLL is received, the ST transmits a DF immediately. A flag bit in the frame header of a DF is set-up when it is the last DF. If the current DF is not the last one and if there is no periodic polling after the end of current DF, the AP sends a CONT that allows the ST to continually transmit one more DF following the current one. BUSY and STOP are used in the same manner as in the periodic polling case.

10.3.7 Multimedia Traffic Support

ATM transport services are divided into several service categories: *constant bit rate* (CBR), *variable bit rate* (VBR), *available bit rate* (ABR), and *unspecified bit rate* (UBR). Even though it is still difficult to perfectly apply the ATM specification to wireless, we will deal with ATM transmission by using the periodic and the non-periodic polling mode.

CBR is very easy to implement; the choice of periodic polling mode with a fixed polling cycle and a fixed-length DF fits the periodic generation of ATM cells. It should be noted that the choice of polling cycle and DF length should be based on BER performance, efficiency and limiting values of the *cell delay variation* (CDV).

In the case of VBR, we can use a periodic polling mode with a fixed polling cycle and a variable-length DF. To fit the traffic variation precisely, it is better to choose a short polling cycle. This choice may, however, result in a degradation of efficiency. Therefore, the polling cycle should be chosen according to mean cell-generation rate, CDV, efficiency and so on.

For ABR, a non-periodic polling mode with a variable polling cycle and a fixed-length DF can be applied. To support dynamic bandwidth control peculiar to ABR service, the ST calculates the target transmission rate according to network congestion level using information from the *resource management* (RM) cells that are fed back from network to the ST, and then requests the AP to adjust the polling cycle to the target rate. In response to the request, the AP adjusts the polling cycle of all STs with ABR in order to maintain fairness of service.

For UBR, the *peak cell rate* (PCR) is the only parameter, that is used during the connection set-up phase. Compared with the other service categories, UBR has the lowest priority and thus, uses non-periodic polling. UBR and other non-real-time services are classified as *best-effort* services, and they can use resource scheduling algorithms for best-effort services [13,14].

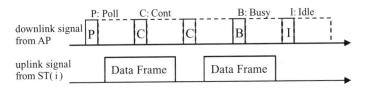

Figure 10.7 Non-periodic polling mode

Although we only described how to support wired ATM network services by RS-ISMA in wireless environments, we believe that the above idea can be applied to other network services such as TCP/IP-based Internet services.

10.3.8 Retransmission Scheme

In RS-ISMA, the QoS requirements of the particular media determine which ARQ scheme is to be used. To guarantee real-time and time-bounded services, a *stop and wait ARQ* (SW-ARQ) with a limited number of retransmissions is used in the periodic polling mode. For non-real-time services, on the other hand, *selective-repeat ARQ* (SR-ARQ) is applied to the non-periodic polling mode.

10.4 BRAIN Indoor LAN Prototype

10.4.1 Configuration

The BRAIN indoor LAN prototype was setup in a furnished office, and the radius of the BSA was about 10 m (Figure 10.8). The AP was hung on the ceiling, and six STs were put

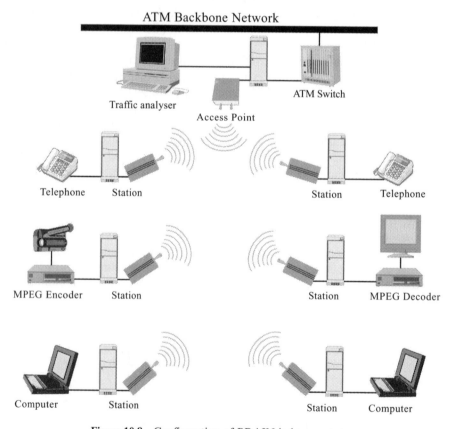

Figure 10.8 Configuration of BRAIN indoor prototype

Table 10.1 Parameters of BRAIN indoor prototype

Frequency	ST: 59.75 GHz AP: 59.25 GHz
Protocols	Multiple access: RS-ISMA Duplex: FDD
Radio trans. rate	51.84 Mbps
Power	ST: 10 mW AP: 15 mW
Modulation	ASK (Burst modem with error correction)
Antenna gain	ST: 20 dBi AP: 5 dBi
Half-power beam width	ST: 15° AP: 60°
Services	Video: MPEG2 Voice: PCM Data: TCP/IP
Network interface	ATM network interface

on different office desks. The AP consisted of physical layer equipment, DLC layer equipment, and a network interface that were connected to both a wired ATM LAN and a traffic analyser. STs consisting of physical layer equipment and DLC layer equipment were connected to video, voice, and data terminals.

Major system parameters of the indoor LAN are listed in Table 10.1. The up-and down-link channels, each with a transmission bit rate of 51.84 Mbps, were carried over separate frequencies in the 60 GHz band. The AP used an antenna with a half-power beam width of 60° in order to cover all the STs each of which had a high-gain directional antenna with a half-power beam width of 15°. *Amplitude shift keying* (ASK) was adopted and implemented in the RF module. The modem supported burst transmission in the uplink and supported BCH code, which improves the BER performance. With regard to the DLC layer, RS-ISMA was implemented in the DLC board. The slot length was designed such that an 80-bit control and management frame was able to be transmitted within a slot.

The system could simultaneously support video (up to 12 Mbps for MPEG II), voice (64 kbps PCM) and data (up to 10 Mbps for *Ethernet LAN emulation* (LANE)) transmissions between wireless STs in the same BSA. LANE-based data communications are available between wireless ST and data terminal connected to the ATM LAN (Figure 10.9).

10.4.2 Implementation

Figure 10.10 shows a block diagram of the major hardware components. The physical layer equipment consisted of a baseband transmitter module, a baseband receiver module

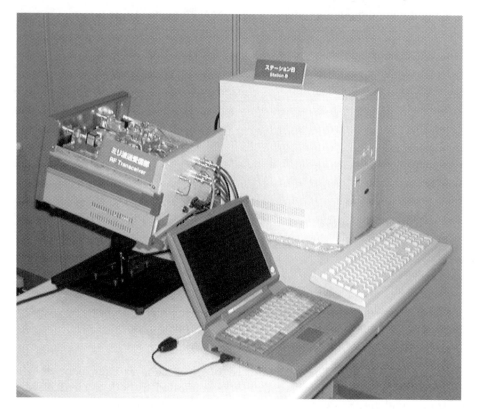

Figure 10.9 Station and data terminal

and an RF module. The DLC layer functions were implemented on a PCI-bus DLC processor board.

10.4.2.1 Physical Layer Equipment

The RF module was composed of transmitter and receiver antennas, and a mm-wave transceiver. The system took advantage of conic horn antennas, which are normally used in mm-wave communications. The mm-wave transceiver used a very simple ASK modulation scheme, i.e. the 60 GHz carrier from Gunn oscillator was switched on/off directly by using a PIN diode. Even though ASK requires a wider bandwidth than those of other modulation schemes, we adopted it because it was much easier to implement due to simplicity and there is sufficient bandwidth at 60 GHz. The transceiver was able to provide a BER of less than 10^{-7} at a bit rate of 51.84 Mbps when there was a 10 m open link between the AP and an ST.

There were a few technical problems concerning the implementation of the very high-speed burst transceiver in the RF modem. The ST had to be able to cut-off transmission power when it was not in the transmit state, not doing so can result in interference with the receiver of AP when another ST is transmitting. Thus, we used two PIN diodes, one for modulating the carrier and one for isolation. The transceiver also required a fast *automatic gain control* (AGC) circuit at the AP receiver because the prototype was a

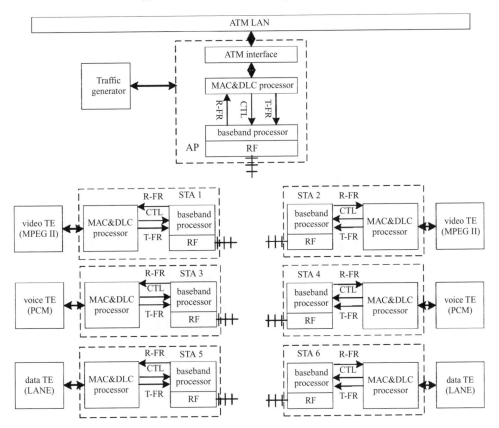

Figure 10.10 Hardware component configuration

packet switching system with a bit rate of over 50 Mbps. In the prototype, the AGC convergent time was reduced to less than 3 μs.

The baseband functions were implemented on a baseband transmitter module and a baseband receiver module. PCI-bus boards were used for each module. After receiving MAC frames to be transmitted from DLC module, the transmitter module scrambled them by 8bit/9bit conversion, encoded them if necessary using BCH(13, 9), and then handed them over to the RF module (Figure 10.11). The receiver module detected the synchronization of physical-layer frames received from the RF module and decoded them.

10.4.2.2 DLC Layer Equipment

The DLC layer functions were implemented on a PCI-bus DLC processor board. The main components of the board were three *field-programmable gate arrays* (FPGA), which had the RS-ISMA protocol downloaded as software (Figure 10.12). The maximum transmission rate in physical layer was 51.84 Mbps; however, the DLC processor board has potential of operating at rates of up to 64 Mbps.

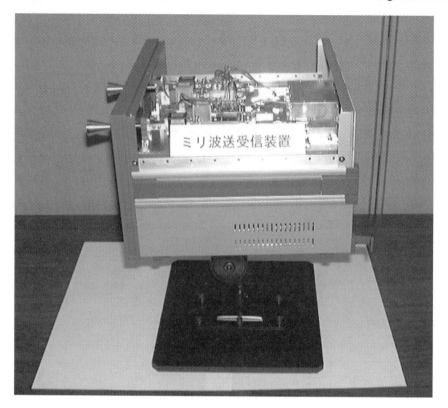

Figure 10.11   RF module

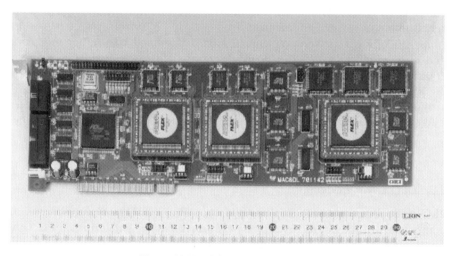

Figure 10.12   DLC processor board

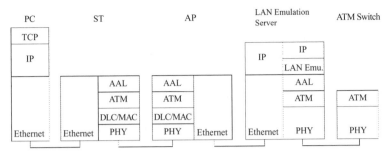

Figure 10.13 Protocol stack for data communication

10.4.2.3 Other Implementations

In the prototype system, an ATM interface was implemented in the AP in order to support data transmission between wireless STs and terminals that were connected to ATM LAN by using LANE. Figure 10.13 shows the protocol stack for this system's data communications. Static IP/MAC address mapping is currently implemented. Dynamic address mapping and mobility management issues are under research.

10.4.2.4 Traffic Analyser

We developed a traffic analyser to check the status of MAC frame stream through the AP and to evaluate the system with various traffic patterns. By connecting the traffic analyser to the AP, we were able to input any speed or pattern of virtual traffic into the system and purposely cause collisions and/or errors in any pattern. The analyser was equipped with *graphical user interface* (GUI) by which we were able to setup these parameters. Furthermore, we were able to monitor traffic via MAC frame error rate, throughput, and offered traffic. This analyser provided enough information to study the system in detail.

References

[1] W. Honcharenko, J. P. Kruys, D. Y. Lee and N. J. Shah, 'Broadband Wireless Access', *IEEE Commun. Mag.*, vol. 35, no. 1, pp. 20–26, 1997.

[2] G. Wu, F. Watanabe, M. Inoue and Y. Hase, 'RS-ISMA with Multimedia Traffic in BRAIN', in *Forty Eighth IEEE Proceedings Vehicular Technology Conference*, Ottawa, Canada, pp. 96–101, May 1998.

[3] K. Kitayama, 'Highly Stabilized Millimeter-wave Generation by using Fibre-optic Frequency-tunable Comb Generator', *IEEE J. of Lightwave Technology*, vol. 15, no. 5, pp. 883–893, 1997.

[4] H. Harada, H. J. Lee, S. Komaki and N. Morinaga, 'Performance Analysis of Fibre-optic Millimeter-waveband Radio Subscriber Loop', *IEICE Trans. Commun.*, vol. E76B, no. 9, pp. 1128–1135, 1993.

[5] K. Sato *et al*, 'Measurements of Reflection and Transmission Characteristics of Interior Structures of Office Building in the 60 GHz Band', *IEEE Trans. on Antennas Propagation*, vol. 45, no. 12, pp. 1783–1792, 1997.

[6] An Introduction to Wireless ATM: Concepts and Challenges. Wireless ATM WG, ATM Forum, 1996.

[7] D. Raychaudhuri and N. Wilson, 'ATM based Transport Architecture for Multiservices Wireless Personal Communication Network', *IEEE J. of Selected Areas in Commun.*, vol. 12, no. 8, pp. 1401–1414, 1994.

[8] P. Agrawal, E. A. Hyden, P. Krzyzanowski, M. B. Srivastava and J. A. Trotter, 'SWAN: A Mobile Multimedia Wireless Network', *IEEE Personal Commun. Mag.*, vol. 3, no. 2, pp. 18–33, 1996.

[9] M. Umehira, M. Nakura, H. Sato and A. Hashimoto, 'ATM Wireless Access for Multimedia: Concept and Architecture', *IEEE Personal Commun.*, vol. 3, no. 5, pp. 39–47, 1996.

[10] G. Wu, Y. Hase, K. Taira and K. Iwasaki, 'A Wireless ATM Oriented MAC Protocol for High-speed Wireless LAN', in *Eighth IEEE Proceedings International Symposium on Personal, Indoor, Mobile Radio Commun.*, Helsinki, Finland, pp. 199–204, Sep. 1997.

[11] G. Wu, K. Mukumoto and A. Fukuda, 'Performance Evaluation of Reserved Idle Signal Multiple-access Scheme for Wireless Communication Networks', *IEEE Trans. on Vehicular Technology*, vol. 43, no. 2, pp. 653–658, 1994.

[12] G. Wu, K. Mukumoto and A. Fukuda, 'Slotted Idle Signal Multiple Access Scheme for Two-way Centralized Wireless Communication Networks', *IEEE Trans. on Vehicular Technology*, vol. 43, no. 2, pp. 345–352, 1994.

[13] M. Inoue, H. Morikawa and M. Mizumachi, 'Size-based Resource Scheduling for Wireless Message Transport', *IEICE Trans. Commun.*, vol. E80-B, no. 3, pp. 466–475, 1977.

[14] M. Inoue, G. Wu and Y. Hase, 'Link-adaptive Resource Scheduling for Wireless Message Transport', *IEEE 1998 Proceedings Global Telecommunications Conference (GLOBECOM98)*, Sydney, Australia, pp. 2223–2228, Nov. 1998.

11

PBX based Mobility Manager for WLL

Yi-Bing Lin

11.1 Introduction

Recently, many telecommunications operators have been looking for wireless technology to replace parts of the hard-wire infrastructure. The *wireless local loop* (WLL) technology [1] is considered as the most fitting solution as radio systems can be rapidly developed, easily extended, and are distance insensitive. Since a WLL eliminates the needs (such as wires, poles and ducts) essential for a wired network, it can significantly speed up the installation process.

A typical WLL system may consist of hundreds or thousands of *base station*s (BSs). In such a large-scale system, the mobility of a user may be limited to a small area. For example, if the *customer premises equipment* (CPE) is a fixed access unit, the user is only allowed to communicate with a specific BS. On the other hand, the user may wish to roam in the whole WLL service area. That is, one may want to connect the handset to any BS in the system. In such a case, the WLL must support mobility management to identify the 'locations' of users. Otherwise, thousands of the BSs would be asked to page a handset for call termination, which is technically infeasible. In this paper, we show how to modify a *private branch exchange* (PBX) to accommodate mobility management for a WLL. The features of our approach are listed below:

1. The PBX will serve as a WLL switch that connects *Public Switched Telephone Network* (PSTN) to the BSs.
2. The BSs are connected to the PBX through standard line or trunk interfaces. We use analogue subscriber lines as an example for PBX-BS connections.
3. The line circuits in the PBX can connect to a wireline telephone or a wireless BS. The PBX automatically distinguishes the BS from the telephone. This feature allows flexible BS layout/installation.
4. A handset can communicate with any BS within the service area of the WLL.

11.2 A Computer-controlled PBX Architecture

Figure 11.1 illustrates a simplified computer-controlled PBX architecture. In this architecture a *Call Processor Switching* (CPX) unit connects to several *Peripheral Modules*

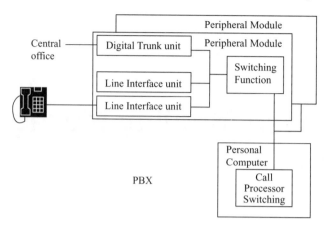

Figure 11.1 A simplified computer-controlled PBX architecture

(PMs). The CPX unit resides in a *personal computer* (PC), which receives and executes commands from the PC, and issues commands to control the PMs. The CPX also performs inter-PM switching functions (i.e. connecting two lines between different PMs). A PM consists of a switching function unit and several telephone interface unit cards. The switching function unit connects every incoming line to the destination outgoing line. The telephone interface unit cards provide various interfaces between the PBX and the outside world. For example, a PM many connect to telephone lines through the *Line Interface Unit* (LIU) or digital trunks through the *Digital Trunk Unit* (DTU). The LIU provides the interface between the PM and a telephone set. The DTU provides the interface between the PM and the trunks connected to the *Central Office* (CO) in the PSTN.

In a PM, every channel (telephone line) in a telephone interface unit card (*or slot*) is associated with a telephone number. This, in PC/CPX, a *telephone table* is required to map the telephone number to a channel. This table is used at the call control process layer to carry out the the call setup and release operations. Following the object-oriented approach, this table can be implemented by a class Telephone Table where every entry in the table consists of three fields (see Figure 11.2(a)): the telephone number, the subscriber profile (to indicate the offered services such as call forwarding, call waiting, and so on), and a pointer to a WireLine object. The WireLine class is used to specify the PM number, the slot number, and the channel number of a telephone line. It also indicates whether the line is busy or idle. Figure 11.2(b) illustrates the physical line configuration that corresponds to the telephone table layout in Figure 11.2(a).

11.3 Mobility Management for PBX

This section describes mobility management in a PBX environment. We assume that the reader is familiar with the concepts of mobility management. General discussions on mobility can be found in [2–5].

Wireless extension to a PBX can be achieved by connecting radio base stations (BSs) to the PM. A BS with large capacity may connect to a DTU (digital trunk unit) card to

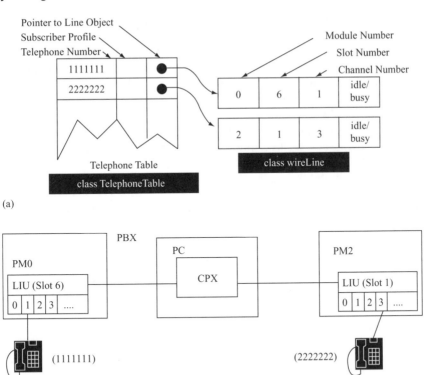

(a)

(b)

Figure 11.2 The telephone table and the corresponding physical configuration: (a) telephone table, and (b) the corresponding physical configuration of (a)

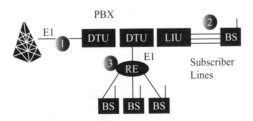

Figure 11.3 Wireless extension to the PBX

provide an E1 link or 30 telephone connections (see (1) in Figure 11.3). A BS with small capacity can connect to LIUs (line interface unit) cards through several subscriber lines (see (2) in Figure 11.3). The small capacity BSs may also connect to a *radio extension* (RE) (e.g. radio port control unit in PACS [6]). The RE then connects to a DTU through the E1 trunk (see Figure 11.3(3)). The PBX configuration described in the previous section must be modified in two aspects to accommodate wireless extension:

Mobility management: If a large number of BSs are connected to the PBX (e.g. 1000–2000 lines), they may be divided into *paging groups*. This concept is similar to the concept of registration area in cellular telephony [3–5]. Location update is required to indicate the paging group where a handset resides.

Handover: When a mobile user is in a conversation, the handset is connected to a BS via a radio link. If the user moves to the coverage area of another BS, the ratio link to the old BS is disconnected and a radio link in the new BS is required to continue the conversation (see Figure 11.4(b)). The radio channel transfer will be taken care by the BSs. During this handover process, the PBX should reconnected the (wire) telephone line from the old BS to the new BS (see Figure 11.4(a)). Three-way calling feature can be used to implement handover at the PBX level. The descriptions are out of the scope of this paper. The reader is referred to Y. B. Lin [7] for more details.

To accommodate mobility management, two modifications to the PBX may be required:

PBX-BS interface. The BS will generate new types of signals to the PBX. There are two alternatives to accommodate these new signal types.

1. The DTU/LIU cards are modified to recognize the new signal types.
2. The BS is modified so that every new signal type is represented by the Off-Hook signal followed by a special code for DTMF signalling or the Seizure signal followed by a special code for trunk signalling [8,9].

We will elaborate the second alternative in this paper.

PBX software. A mobility management software should be created, and minor modifications to the call control process are required to implement mobility management.

We use the subscriber line connection (see Figure 11.3(2)) to illustrate the implementation of wireless extension to the PBX.

For wireless extension based on the subscriber line connection (see Figure 11.5), a BS connects to the PBX through several subscriber lines. For a call originated from a handset in the BS coverage, an arbitrary idle subscriber line is selected by the BS to connect the PBX. For a call termination to a handset, the PBX selects an arbitrary idle line to the BS. We make the following assumptions.

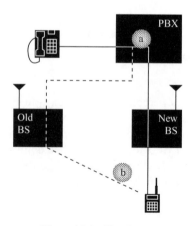

Figure 11.4   Handover

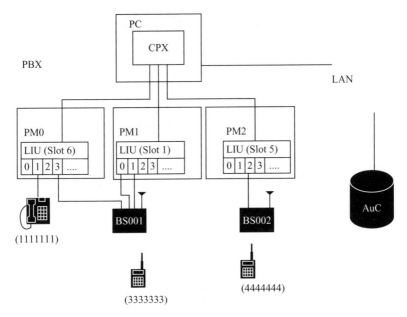

Figure 11.5 An example of wireless Extension to the PBX

Assumption 1.   Every paging group consists of one BS. In other words, location update is performed every time a handset moves from a BS to another. For every call termination to a handset, at most one BS is asked to page the handset. We note that it is trivial to relax this assumption to accommodate multiple BSs in a paging group.

Assumption 2.   Every BS has c_1 radio channels and c_2 wireline connections to the PBX where $c_1 \geq c_2$. If the BS is capable of handling intra-BS calls, then $c_1 > c_2$ in general. An intra-BS call does not need connection to the PBX as illustrated in Figure 11.6(a). If the BS cannot connect intra-BS calls by itself then the call is handled like an inter-BS call as illustrated in Figure 11.6(b). In this case $c_1 = c_2$. In our approach, a BS may inform the PBX whether it can handle intra-BS calls or not.

Assumption 3.   An authentication centre (AuC) is required to authenticate the handsets. The AuC may or may not collocate with the PBX. The handset is authenticated via its password. This paper assumes that the AuC is implemented within the PBX.

Compared to the configuration in Figure 11.2(b), the configuration in Figure 11.5 introduces several new entities: handset (mobile phone), base station (BS), and *authentication centre* (AuC). The implementation for these entities are described below.

Line class.   An abstract Line class is introduced. From this abstract class, two classes WireLine and WirelessLine are derived as illustrated in Figure 11.7. The WireLine class is the same as before except that an extra BS pointer is included (see Figure 11.8(a)). For a subscriber line connected to the wireline telephone, the BS pointer value is NULL. If a subscriber line is connected to a BS, then the corresponding WireLine object will point to the BaseStation object of the connected BS. Every WirelessLine object is associated with a handset. This class consists of three fields (Figure 11.8(f)): a status bit to indicate if the handset is busy, an AuC address (pointer to an authentication table where the handset's password is stored), as BS address (pointer to a BS object corresponding to the BS coverage where the handset resides).

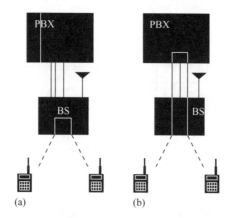

Figure 11.6   Intra-BS call connection: (a) BS with intra-BS switching ability, and (b) BS without intra-BS switching ability

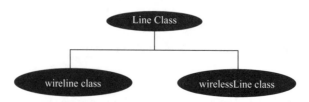

Figure 11.7   The Line class hierarchy

BaseStation class.   A BaseStation object maintains a circular linked list of pointers to the WireLine objects (Figure 11.8(b)). These WireLine objects correspond to the subscriber lines connecting the BS and the PBX. The BaseStation object also has a record to store the BS profile (encryption information, intra-BS call ability, and so on).

BS Table class.   The BSTable object (Figure 11.8(c)) maintains the list of BSs connected to the PBX.

AuCTable class.   The AuCTable object (Figure 11.8(d)) is a table that stores the authentication keys (passwords) of the handsets in the system. The authentication procedure is out of the scope of this paper but can be found in [3,10,11].

11.4   Registration Procedures

When the PBX subscriber lines are first connected to a BS, BS line registrations are required (one registration per line). Consider BS001 in Figure 11.9. After channel 3 of Slot 6 in PM0 is connected, BS001 initiates the registration procedure through subscriber line signalling as illustrated in Figure 11.9(a). The BS line registration procedure is described in the following steps:

Steps 1 *and* 2. BS001 sends an Off-Hook signal to the PBX. The PBX replies dial tone to BS001, and expect to receive DTMF digits from BS001. Since these two steps are exact the same as that in the subscriber line call set-up procedure, they are handled by the

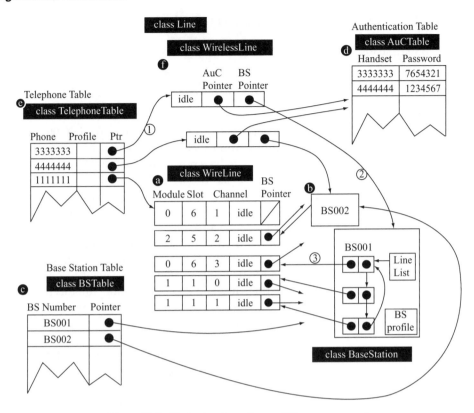

Figure 11.8 Data structures for mobility management

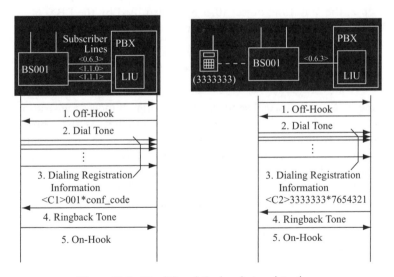

Figure 11.9 The BS and the handset registrations

normal PBX call control process. At this point, a WireLine object $\langle 0, 6, 3 \rangle$ is created (Figure 11.8(a)).

Step 3. BS001 sends a sequence of DTMF digits

$$\langle C1 \rangle 001 * conf\_code$$

to the PBX where $\langle C1 \rangle$ is a special sequence representing BS line registration, 001 is the BS number, and *conf_code* provides BS information such as the ability of handling intra-BS calls. This sequence is analysed by the Digit-Analysis procedure used in the normal PBX call control process. By detecting the code $\langle C1 \rangle$, the BSTable (Figure 11.8(c)) is searched to locate the BaseStation object (Figure 11.8(b)) for BS001. If the BS001 entry is not found in the table, a new entry and a new BaseStation object BS001 is created for BS001. The WireLine object $\langle 0, 6, 3 \rangle$ (Figure 11.8(a)) is added to the LineList of BS001, and the BS pointer field of the $\langle 0, 6, 3 \rangle$ WireLine object is set to the address of BS001.

Steps 4 *and* 5. After the PBX has processed the DTMF digit sequence, it sends a tone signal (or a DTMF signal) to BS001. BS001 sends an On-Hook signal to complete the registration process. Note that if the PBX needs more information for the registration process, it may send a different DTMF signal to BS001. Based on the DTMF signal, BS001 may resend the DTMF string or abort the registration action.

The above procedure repeats for every subscriber line connected to BS001. For the data structures configuration in Figure 11.8, the corresponding physical layout is shown in Figure 11.5.

When a handset 3333333 arrives at BS001, a handset registration is required. The registration message flow is illustrated in Figure 11.9(b), and the steps are described below:

Steps 1 *and* 2. These steps are the same as that for BS line registration. BS001 selects an idle subscriber line for signalling. Suppose that the line is channel 3 of Slot 6 in PM0. At the end of Step 2, the WireLine object $\langle 0, 6, 3 \rangle$ is identified by the PBX.

Step 3. BS001 sends the DTMF digit sequence

$$\langle C2 \rangle 3333333 * 7654321$$

to the PBX where $\langle C2 \rangle$ is a special sequence representing handset registration, 3333333 is the handset number, and 7654321 is the password. The Digit-Analysis procedure detects handset registration from the special code $\langle C2 \rangle$ and searches the phone entry 3333333 in the TelephoneTable (Figure 11.8(e)). We note that all legal handsets are recorded in the TelephoneTable and the corresponding WirelessLine objects (Figure 11.8(f)) are created through an off-line procedure at system initialization. Thus, the telephone table entry is always found for a legal handset registration. Through the WirelessLine object, the handset is authenticated by using the number 7654321 in the DTMF string and the password stored in the AuCTable (Figure 11.8(d)). If the authentication process is successful, the BS pointer of the WirelessLine object is assigned the BS pointer value of the WireLine object $\langle 0, 6, 3 \rangle$.

Steps 4 *and* 5. These steps are the same as Steps 4 and 5 in the BS line registration procedure.

11.5   Call Termination

Suppose that the wireline phone 1111111 calls the handset 3333333. The signalling procedure is illustrated in Figure 11.10.

Steps 1–3. These three steps are similar to that in the registration procedures in Figure 11.9. The difference is that the PBX recognizes the dialed DTMF digits as a phone number. The TelephoneTable object is searched to locate the WirelessLine object of 3333333 (see path (1) in Figure 11.11). If the search indicates that the handset is idle, then the PBX sets both the calling and the called Line objects 'busy' and locates the BaseStation object of the BS (i.e. BS001) where the handset resides (see path (2) in Figure 11.11). The PBX searches LineList of BS001 to find the first idle channel ($\langle 0, 6, 3 \rangle$ in this example; see path (3), Figure 11.11).

Steps 4–6. Through the $\langle 0, 6, 3 \rangle$ WireLine object, the PBX informs BS001 of the call termination to the handset 3333333. Steps 4–6 are similar to Steps 1–3. At the end of Step 6, BS001 pages the handset using the received handset number.

Step 7. If the handset (not the user of the handset) responds to the page, an Off-Hook signal (similar to the SS7 ISUP ACM message [12]) is sent from the BS to the PBX.

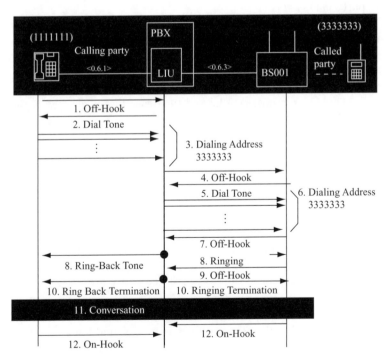

Figure 11.10   The wireline to wireless call

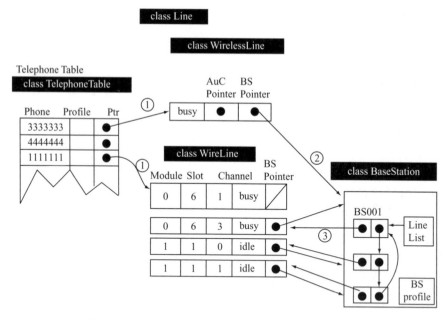

Figure 11.11 Data structure configuration for wireline-to-wireless call

Step 8. The PBX sends a ringing tone to the called party (BS001), and a ring-back tone to the calling party (1111111).

Step 9. If the user of the handset 3333333 answers the phone, a second Off-Hook signal is sent from BS001 top the PBX.

Step 10. The PBX detects the called party Off-Hook signal, and then removes the ringing and the ring-back tones.

Note that if the call set-up procedure fails for some reason, the Line objects correspond to 3333333 and 1111111 are marked 'idle' and the procedure is aborted. Otherwise, these objects are marked 'busy' at Step 3, and the procedure proceeds to Step 11.

Step 11. The voice path is connected and both parties start conversation.

Step 12. When either party hangs on, the On-Hook signal is sent to the PBX and the connection is released (the WirelessLine object for 3333333, the WireLine objects $\langle 0, 6, 1 \rangle$ and $\langle 0, 6, 3 \rangle$ are set 'idle').

Consider an intra-BS call where the calling handset 3333333 and the called handset 4444444 are both in the radio coverage area of BS001. If the BS cannot switch the intra-BS call, then the signalling and switching procedure is exact the same as that in Figure 11.10. If the BS can handle the intra-BS call without subscriber line setup through the switch, then the signalling procedure is illustrated in Figure 11.12.

Steps 1–3. These three steps are similar to that in the handset registration procedure in Figure 11.9(b). BS001 selects $\langle 0, 6, 3 \rangle$ for all origination signalling to the PBX. The PBX marks the WireLine object $\langle 0, 6, 3 \rangle$ 'busy' and recognizes the dialed DTMF digits as a wireless handset call origination (represented by the special code $\langle C3 \rangle$) where the calling

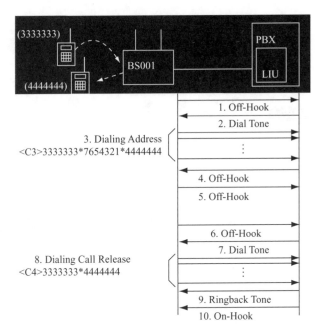

Figure 11.12 The intra-BS call signalling procedure

handset number is 3333333, its password is 7654321, and the called party number is 4444444. The TelephoneTable is searched to locate the WirelessLine object of 3333333 (see path (1) in Figure 11.13). Suppose that the line object indicates the idle status. The status is marked 'busy'. The BS pointer of the object is used to compare with the BS pointer of the WireLine object $\langle 0, 6, 3 \rangle$. If they do not match, then an error is detected. This situation occurs when a handset makes a call at a BS before it registers to that BS. The authentication procedure is performed as described in the registration procedure (see path (2) in Figure 11.13). Then the Line object of the called handset is located as described at Step 3 of the wireline-to-wireless call procedure (see path (3), Figure 11.13). The PBX realizes that both the calling and the called parties are in BS001.

Steps 4 *and* 5. Through the same subscriber line $\langle 0, 6, 3 \rangle$, the PBX informs the base station BS001 of an intra-BS call using the Off-Hook signal. BS001 pages 4444444, and connects 3333333 to 4444444 directly. BS001 informs the PBX that the call is set up by an Off-Hook signal. At this point, both Line objects for 3333333 and 4444444 are 'busy'. The PBX sets $\langle 0, 6, 3 \rangle$ 'idle'.

Step 6. The conversation begins without using any subscriber line between BS0001 and the PBX.

Steps 7–10. At the end of the conversation, BS001 releases the radio channels, and sends a DTMF sequence with the special code $\langle C4 \rangle$ to the PBX. The code represents 'end of intra-BS call'. Based on the handset numbers (3333333 and 4444444), the PBX marks the corresponding WirelessLine objects 'idle', which completes the intra-BS call transaction.

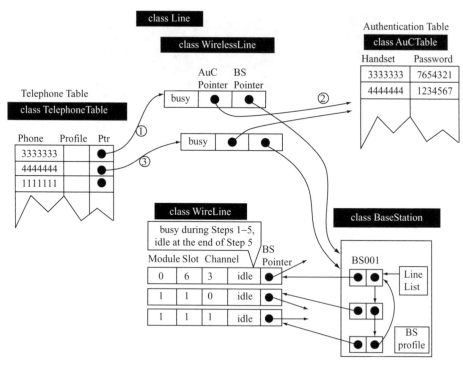

10.13   Data structure configuration for intra-BS call

11.6   Performance Issues

Signalling traffic due to mobility may degrade the performance of the PBX. Our experience with the computer-controlled PBX is that the extra CPU processing cost for mobile operations (specifically, registrations and wireless call terminations) are negligible compared with the normal call-set-up processing cost.

A potential problem occurs when handset registration utilizes a subscriber line from the BS to the PBX. If all subscriber lines are busy when a registration request arrives, the system may simply drop the registration. On the other hand, we can still allow registration operation by sending DTMF signalling through a busy line. In this case, the user of the subscriber line will hear the DTMF tone during the conversation. Such DTMF 'noise' may be considered as degradation of circuit quality. Thus, it is important to see how often a busy line is interrupted by the registration operations. We propose an analytic model to investigate this issue. Suppose that the call arrivals are Poisson process with rate λ, and the call holding times are exponentially distributed with mean $1/\mu$. If there are c subscriber lines between the BS and the PBX, then the probability p_b that all subscriber lines are busy can be expressed by the Erlang B formula

$$p_b = \frac{(\rho^c/c!)}{\sum_{0 \leq j \leq c}(\rho^j/j!)}$$

where $\rho = \lambda/\mu$. Suppose that the registration traffic forms a Poisson process with rate γ. The rate of registration traffic that will interrupt a particular line (with probability $1/c$) is thus

$$\theta = \frac{\gamma p_b}{c} \tag{11.1}$$

From Equation (11.1), the number k of registrations that will affect a call during its call holding time t_h has a Poisson distribution with density function

$$f_k(t_h) = \frac{(\theta t_h)^k}{k!} e^{-\theta t_h}$$

The probability $Pr[K = k]$ that there are k registration interruptions to a call is

$$
\begin{aligned}
Pr[K = k] &= \int_{t_h = 0}^{\infty} f_k(t_h)\mu e^{-\mu t_h} dt_h \\
&= \left(\frac{\theta^k}{k!}\right) \int_{t_h=0}^{\infty} t_h^k \mu e^{-(\theta+\mu)t_h} dt_h \\
&= \left(\frac{\theta^k}{k!}\right) \left[\frac{\mu k!}{(\theta+\mu)^{k+1}}\right] = \frac{\mu \theta^k}{(\theta+\mu)^{k+1}}
\end{aligned}
\tag{11.2}
$$

From Equation (11.2), the expected number $E[K]$ of registration interruption to a call is

$$E[K] = \sum_{i=1}^{\infty} i Pr[K = i] = \frac{\gamma p_b}{c\mu + \gamma p_b} \tag{11.3}$$

Table 11.1 lists the number $E[K]$ of registration interruptions to a call for various call and registration traffics where $c = 8$. Suppose that the mean call holding time is $1/\mu = 3\,\text{min}$. For the worst case in the table (where the call arrival rate is 2.67 calls per minute and the registration rate is 0.5 per minute), the probability p_b of no idle circuits is 23.56 % and the expected number of interruptions to a call is less than 0.05. The above analysis indicates that under reasonable conditions, the PBX calls are seldom affected by the mobility traffic.

Table 11.1 The number of registration interruptions to a call ($c = 8$)

λ/μ	4.0	5.0	6.0	7.0	8.0
p_b (%)	3.04	7.00	12.19	17.88	23.56
$E[K]$ ($\gamma/\mu = 0.5$)	0.001898	0.004359	0.007560	0.011053	0.014510
$E[K]$ ($\gamma/\mu = 1.0$)	0.003788	0.008680	0.015006	0.021864	0.028604
$E[K]$ ($\gamma/\mu = 1.5$)	0.005671	0.012964	0.022341	0.032441	0.042301

References

[1] D. C. Cox, 'Wireless Loops: What are They?' *International J. Wireless Information Networks*, vol. 3, no. 3, pp. 139–145, 1996.

[2] Y.-B. Lin, 'Mobility Management for Cellular Telephony Networks,' *IEEE Parallel Distributed Technology*, vol. 4, no. 4, pp. 65–73, 1996.

[3] M. Mouly and M. B. Pautet, *The GSM System for Mobile Communications. 49 rue Louise Bruneau*, Palaiseau, France, 1992.

[4] EIA/TIA. Cellular intersystem operations (Rev. C), *Technical Report IS-41, EIAiTIA*, 1995.

[5] ETSI/TC. Mobile application part (MAP) specification. Version 4.8.0. Technical Report Recommendation GSM 09.02, ETSL, 1994.

[6] A. R. Noerpel, Y. B. Lin and H. Sherry, 'PACS: Personal Access Communications System—A Tutorial,' *IEEE Personal Commun. Mag.*, pp. 32–43, 1996.

[7] Y. B. Lin, 'PACS Network Signaling using AIN/ISDN,' *IEEE Personal Commun. Mag.*, vol. 4, no. 3, pp. 33–39, 1997.

[8] R. F. Rey, *Engineering and Operations in the Bell System*, AT & T Bell Laboratories, 1989.

[9] J. H. Green, *The Irwin Handbook of Telecommunications*, Pantel Inc., 1997.

[10] EIA/TIA. Cellular Radio-Telecommunications Intersystem Operations: Authentication, Signaling Message, Encryption and Voice Privacy, *Technical Report TSB-51. EIA/TIA*, 1993.

[11] S. Redl, M. Weber, *An Introduction to GSM*, Artech House, London, 1995.

[12] ANSI. American National Standard for Telecommunications—Signaling System Number 7 Integrated Services Digital Network (ISDN) user part, Issue 2, Rev. 2. *Technical Report ANSI T1.113, ANSI*, 1992.

12

Remote Management and Upgrade in a Wireless Local Loop System

Thomas Jagodits and Hans Bhatia

12.1 Introduction

Wireless Local Loop (WLL) is projected to experience phenomenal growth over the next years. The nature of WLL is such that terminals are not mobile and once installed the subscriber expects trouble-free operation without any service requirements. Meanwhile, new functionality involving enhanced codecs and modems are being developed and expected to be made available to the subscriber. A major challenge facing WLL operators consists of upgrading WLL terminals that are already deployed at the customer site with new functionality [1,3]. Another challenge is the ability to monitor terminal service parameters and events without obtaining physical access to the device.

The main factor that enables scalable propagation of software images and tracking events at WLL terminals is the dissemination of FLASH-based devices. The cost of FLASH memory has been decreasing, becoming the memory device of choice in many wireless devices.

This chapter presents mechanisms that enable propagation of software images to wireless terminals and remote capture of events on the same terminals. Both mechanisms have been implemented in a Wireless Local Loop system. The remote upgrade mechanism was used to successfully propagate dozens of software releases in multiple markets. The monitoring mechanism is being used daily to remotely monitor the operational status of thousands of WLL terminals.

Section 12.2 provides a general overview of the WLL application, Section 12.3 details the remote upgrade mechanism including implementation at the wireless device, in this case a Subscriber Unit. Section 12.4 describes the event logging mechanism used for remote management and the last section concludes this chapter.

12.2   Wireless Local Loop Application Overview

Wireless Local Loop (WLL) provides affordable local telephony access that can be deployed quickly. Several companies provide WLL solutions, the most popular digital air-interface protocols being *Time Division Multiple Access* (TDMA) and *Code Division Multiple Access* (CDMA). Hughes Network Systems AIReach(tm) Local Loop TDMA system provides a flexible, high-capacity WLL platform based on digital TDMA access. This system is currently deployed in 8 countries with over one million subscribers under contract worldwide.

In order to propagate new features and software fixes to the installed base a mechanism that involves over the air propagation (remote upgrade) of software images was implemented in the *Hughes Network Systems* (HNS) platform. To enable this mechanism, HNS expanded the air-interface protocol and implemented collateral modifications at both the *Base Station Controller* (BSC) and the *Fixed Subscriber Unit*s (FSUs). The FSU is a WLL terminal located at the subscriber premises, providing residential or commercial wireless access. Figure 12.1 shows a typical WLL system.

By definition FSUs cannot move into a better RF environment to alleviate signal fading. Subscribers also expect them to operate reliably with little or no maintenance. In a commercial system with tens or hundreds of thousands of units the need to remotely gather operational information from each unit is critical. FSUs in the HNS system continuously capture operational and link-based events. These data are relayed on demand to the BSC where it is used to maintain overall system statistics and take corrective actions when necessary.

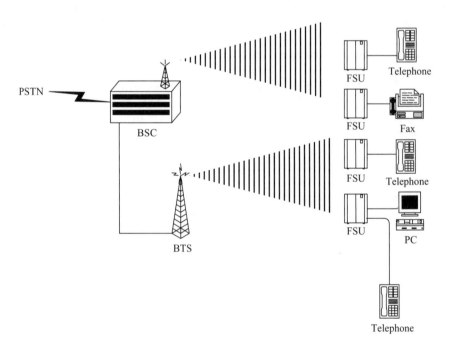

Figure 12.1   Typical WLL System

12.3   Remote Upgrade of Fixed Subscriber Units

12.3.1   Characteristics

The key functional characteristics of the remote upgrade or *down-line-load* (DLL) mechanism are:

Broadcast   By the virtue of the wireless transmission medium, broadcast is used for simultaneous download to all or selected subscriber units. The transmission is secure to the extent that the receiving subscriber unit complies with the FE-TDMA (*Fixed Enhanced TDMA*) and down-line-load protocols to receive and decode messages.

Retransmission   The subscriber units may also inform the BSC about missing records after each load module has been transmitted. The reporting period is randomized among subscriber units to avoid collision on the same transmission medium, which rarely should happen due to the fact that missing records event itself occurs randomly among the FSUs. The BSC collates the missed records and retransmits them as a bundle at the end for optimal throughput.

Automated Download   While network load is low, the BSC scans through the database of registered subscriber units and automatically downloads a configured release to units operating with older releases.

Software Image Partitioning   The whole software image to be down line loaded is labelled and partitioned into multiple load modules. Each one of these load modules is transmitted over the air in an order specified by the BSC.The remote upgrade (DLL) is initiated at the BSC via a *graphical user interface* (GUI) or *command line interface* (CLI). The minimum parameters required for a DLL are the label of the new software image and the specification of the cell in which the down-line-load is to be performed. Optional parameters include *Mobile Identification Number*s (MINs) and the specifying the label of the existing software to be upgraded or downgraded. These last two parameters are exclusive.

Each release consists of multiple load modules and three configuration files. The load modules contain the data to be broadcast by the BSC. The first configuration file lists the files to be transmitted by the BSC when a release is downloaded. The second configuration file contains the checksum of the load modules, it is used by the BSC to verify the integrity of the load module files before transmitting them. The third configuration file determines how to load an FSU with a release. This file specifies how to upgrade an FSU to a new release and how to downgrade an FSU to a previous release. In some cases, it might be necessary to first load an intermediate release before loading the final release. This configuration file informs the BSC of the upgrade path.

When the BSC receives the download command from the Operator Console or as a part of the automated download, the BSC filters out the list of subscriber units to be downloaded. Then it selects a list of free channel slots, allocates them for download and broadcasts this information through a *Broadcast Forward Control* (BFC) message to all the subscriber units to prepare the units to receive information conveyed over the download channels.

After the BFC message, the Base Station Controller will send a DLL Header message on all the download channels. The subscriber units use this information to decide whether to participate in the DLL (see Figure 12.2). Next the BSC proceeds to download each individual image file sequentially. Subsequently the BSC checks if it can access the load module file stored on disk of the Operator console by mounting the corresponding file system via the backbone LAN. The file is transferred through the standard File Transfer Protocol [2] to local memory. If the transfer is successful, the transmission of load modules over the air is initiated on a channel.

Missed records reported by subscriber units are assembled at the BSC. At the end of the transmission of the load module, the BSC explicitly requests the subscriber units to inform about the missed records. After collating all the responses, it retransmits those that were missed. This process is repeated until no responses are received for the retransmission

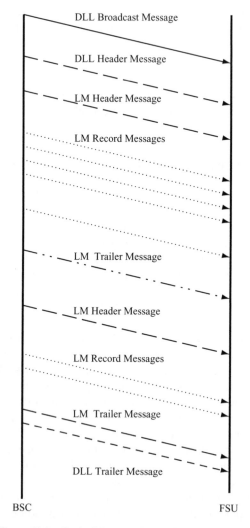

Figure 12.2 Typical Remote Upgrade Message Flow

request, at which time the BSC sends Load Module Trailer message to the subscriber units for indication of the completion of load module and for checksum match.

The process of load module transmission is repeated for all the load modules that are part of an image file. At the end, the BSC sends the DLL complete trailer message to the subscriber units and deallocates the channels used for the DLL. A remote upgrade may be performed to specific MINs, specific FSU models or to a whole cell. It is initiated through the BFC message. The messages that comprise the actual down-line-load are: DLL header message, DLL Trailer Message, Load Module Header Message, Load Module Record and Load Module Trailer Message.

The contents of each message are as follows:

- *DLL Header Message*—This message is the very first message transmitted by the BSC when starting the DLL. This message is transmitted on a channel selected for the down-line-load channel by the BSC. It contains all the information necessary by the FSUs to decide if they will participate in the down-line-load process.
- *DLL Trailer Message*—This message is the last message transmitted by the BSC, it contains the status of the DLL termination. If the DLL is aborted at the BSC it will be indicated in this message. This message can be send at any time during the down-line-load by the BSC.
- *Load Module Header Message*—This message which is send by the BSC is the preamble for individual load modules. It contains information on the overall load module size and the number of records in this load module.
- *Load Module Record*—Comprises the actual image data. In addition it also contains a sequence number so any message loss can be detected immediately. Each Load Module typically consists of multiple load module records.
- *Load Module Trailer Message*—Message send by the BSC indicating the end of transmission of a load module. It contains a checksum that enables the immediate detection of any corruption in the load module. It provides a second level of integrity on the data, the modem provides the CRC check on individual messages, this checksum provides the CRC check on the whole load module.
- *Missing Record Request Message*—Message send by the BSC that can be used for retransmission of specific DLL records. It contains data and sequence numbers of the records that are being retransmitted.
- *Missing Record Response Message*—this is the message that optionally can be sent by FSUs to request missed records. It contains a list of records that were not received by the FSU.

12.3.2 Remote Upgrade Processing at the Fixed Subscriber Unit

A DLL is initiated at the BSC by transmitting the corresponding BFC message. The Fixed Subscriber Unit processes the DLL notification by tuning to the first DLL stream.

On receipt of a DLL Header message the Fixed Subscriber Unit will ensure that this is not a duplicate message and if the unit is in a state in which it can accept a down-line-load. For small scale FSUs that typically are located in residential premises, this involves checking if any calls are up or if the unit is running on battery power. The FSU will also verify if the unit should participate in the download, this is achieved by processing the image label and MIN fields. If the header indicates that the down-line-load is targeted for a different FSU model the unit will not participate in the DLL. Similarly, if the unit already has the software image loaded or if it's MIN is not in the provided MIN list the unit will simply stop tuning to the DLL stream.

On receipt of the first Load Module Header message the FSU will allocate memory to process the load module records. After a complete load module is received, that is, all load module records have been received, the load module is stored in a pre-defined scratch area in FLASH. A load module is only moved to the actual target image location when all load modules and the appropriate DLL trailer message have been received. In this manner the unit is not impacted if there is a power outage or if individual load modules are corrupted. If the FSU is participating of a DLL and receives the DLL Trailer message with an indication that the DLL has been completed successfully at the BSC it confirms that the load modules have been received properly by performing the checksum on all load modules. If the checksum matches it will update the FLASH with the new image and restart the unit.

A DLL is aborted if any of the following three conditions occur. If the DLL Trailer message indicates that the DLL should be aborted the DLL is terminated. And also if the subscriber originates or receives a call the DLL is terminated, this ensures transparency to the end user. Depending on the model it is desirable to abort the download if the unit is operating on battery because writing to FLASH in very low power conditions could corrupt the memory.

The option to selectively request missing records is not always exercised at the subscriber unit. If too many subscriber units would request missing records simultaneously it could impact the operation at the BSC. Instead the software upgrade relies on the automated download capability to ensure that all terminals are upgraded.

12.4   Event Logging for Remote Management

During operation the FSU captures link and operational events continuously. These are selectively stored in RAM or FLASH memory and include forward channel based RF statistics, hardware failures if any, operational inconsistencies, power events, etc.

The remote management is initiated periodically at the BSC through a screen or through a command line interface script when network traffic is low. Optionally remote management can be started at the BSC for monitoring purposes or in response to specific customer complaints.

The remote management implementation in the HNS system involves three distinct operations necessary to provide remote network management: link, terminal and poll. Each of these operations consists of a query by the BSC and response by the FSU.

Remote Management Link Operation   To provide link data the FSU continuously gathers forward RF information over time. It maintains statistics on minimum and maximum RSSI values, BER, CRC errors and number of messages received per frequency. Each message received from the BSC in the forward direction is an event and the corresponding information is stored in RAM. The operation query is received through a MIN based FC message and the response is relayed through FACCH messages, the BSC will retry the operation if no response is received from the FSU.

Remote Management Terminal Operation Information pertaining to this operation includes state of the FSU, reason and time of the last resets, forward signal quality indication, power status and status of the last down line load. With the exception of the signal quality indication and power status all data is stored in FLASH each one as a distinct event with corresponding timestamp. These event data are not lost between resets. The operation query by the BSC is also received through a MIN based FC message and the response is relayed through FACCH messages, the BSC will retry the operation if no response is received from the FSU.

Poll Operation This consists of a query with minimal overhead to verify if the FSU is operational and is similar to the IP based 'ping' command. This command consists of a FC Poll query message and response; it requires minimal air bandwidth and typically is not retried at the BSC.

Output of the operations is saved in files at the BSC for post processing or future reference. This provides the ability for tracking the wireless system over time, comparing distinct WLL networks and identifying any performance issues. Table 12.1 shows a sample output of a terminal operation for seven FSUs.

Similarly the results of the link operation are MIN based. Table 12.2 shows a sample output for a two FSUs in a cell with 12 frequencies. The second unit sampled multiple messages with errors on several frequencies, it also sampled from both antennae.

If in the future additional operational information has to be relayed from the FSU to the BSC this can be achieved by simply upgrading all the FSUs in the WLL network with a new release which provides the necessary data using the remote upgrade mechanism.

Presently commands and responses exchanged between the BSC and the FSUs are packed to minimize bandwidth usage. All network management commands, responses and user interfaces are crafted according to their specific purpose. A future optimization, which is becoming a trend for embedded devices, consists in including web server capabilities at the FSUs. In this manner no modifications would be required at the BSC when additional operational data are relayed by the FSUs.

Table 12.1 Sample terminal output

MIN	Cell	Sector	Registration Date	SW	Upgrade Status
0907202721	TURBHEWTX	SECTOR3	06–Feb 1999 04:32:25	s07408 1 4 42 OK N	DLL_SUCCESSFUL
0907234507	PARSIKHILL	SECTOR3	06–Feb 1999 04:31:28	s07408 2 3 72 OK N	DLL_SUCCESSFUL
0907243865	TURBHEWTX	SECTOR3	06–Feb 1999 04:31:41	s07408 1 4 40 OK N	DLL_SUCCESSFUL
0907245631	AIROLI	SECTOR2	05–Feb 1999 01:25:18	s07408 3 3 64 OK Y	SSU_REL_NOT_EQ_OLD
0907247155	AIROLI	SECTOR2	06–Feb 1999 04:28:57	s07408 2 3 60 OK N	SUBS_OFF_HOOK
0907247175	PARSIKHILL	SECTOR1	06–Feb 1999 04:31:25	s07408 1 3 66 OK N	DLL_SUCCESSFUL
0907247211	PARSIKHILL	SECTOR3	06–Feb 1999 04:30:59	s07408 3 3 63 OK N	DLL_SUCCESSFUL

Table 12.2 Sample link operation output

MIN	Fr	Good Message	Sync Errors	CRC Errors	BER	WER%	CI Mn	CI Mx	CI Av	AntA Min	AntA Max	AntA Avg	AntB Min	AntB Max	AntB Avg
0907274502	1006	23576	0	0	0.00	0.00	9	21	16	-85	-80	-82	0	0	0
0907274502	181	23574	0	0	0.00	0.00	10	22	16	-87	-81	-83	0	0	0
0907274502	199	23577	0	0	0.00	0.00	7	21	14	-89	-83	-85	0	0	0
0907274502	145	23580	0	1	0.02	0.00	3	22	13	-98	-85	-89	0	0	0
0907274502	163	23577	0	0	0.00	0.00	4	22	16	-87	-81	-83	0	0	0
0907274502	289	23575	0	0	0.00	0.00	8	22	14	-90	-83	-86	0	0	0
0907274502	319	23577	0	0	0.01	0.00	3	23	15	-108	-87	-92	0	0	0
0907274502	271	23575	0	0	0.00	0.00	6	21	16	-93	-84	-86	0	0	0
0907274502	307	23577	0	0	0.00	0.00	7	20	14	-92	-84	-87	0	0	0
0907274502	325	23577	0	0	0.00	0.00	6	21	14	-93	-84	-87	0	0	0
0907274502	235	23577	0	0	0.00	0.00	9	20	16	-86	-81	-83	0	0	0
0907274502	217	23575	0	0	0.00	0.00	7	21	14	-91	-83	-85	0	0	0
0907274600	1006	14406	0	0	0.00	0.00	9	24	15	-86	-77	-80	-94	-81	-88

0907274600	181	18330	1	1	0.00	0.01	0	23	16	−113	−79	−84	−100	−85	−90
0907274600	199	14053	667	161	0.07	1.13	0	24	13	−105	−81	−86	−119	−89	−97
0907274600	145	27167	0	9	0.00	0.03	4	22	15	−112	−83	−91	−95	−84	−89
0907274600	163	19495	0	0	0.00	0.00	9	21	15	−84	−76	−79	−86	−80	−82
0907274600	289	27891	0	0	0.00	0.00	8	21	15	−88	−79	−82	−96	−82	−87
0907274600	319	14351	0	0	0.00	0.00	6	22	15	−84	−77	−79	−100	−86	−90
0907274600	271	18328	0	0	0.00	0.00	10	21	16	−86	−78	−80	−89	−81	−84
0907274600	307	14196	334	6	0.01	0.04	0	22	13	−90	−79	−83	−109	−89	−95
0907274600	325	17614	0	0	0.00	0.00	7	22	13	−85	−78	−80	−93	−84	−87
0907274600	235	17560	0	0	0.00	0.00	9	21	15	−89	−79	−81	−91	−83	−86
0907274600	217	18157	0	0	0.00	0.00	6	21	14	−90	−79	−82	−93	−82	−85

References

[1] J. Mitola III. 'Technical Challenges in the Globalization of Software Radio,' *IEEE Commun. Mag.*, vol. 37, no. 2, pp. 84–89, 1999.

[2] J. Postel, J. Reynolds. *File Transfer Protocol, RFC 959*, 1985.

[3] J. Ryan. 'Interfacing Converters and DSPs,' *Communication Systems Design*, vol. 5, no. 3, 18–27, 1999.

13

Current and Future Services Using Wireless Local Loop (WLL) Systems

Dong Geun Jeong and Wha Sook Jeon

13.1 Introduction

In the telephone networks, the circuit between the subscriber's equipment (e.g. telephone set) and the local exchange in the central office is called the 'subscriber loop' or 'local loop'. Traditionally, the copper wire has been used as the medium for local loop to provide voice and voice-band data services. Since 1980's, the demand for communications services has increased explosively. There has been a great need for the basic telephone service, i.e. the *plain old telephone service* (POTS) in developing countries. On the other hand, in the industrialized countries, the demand for high-speed data and multimedia services at home and/or office has increased continuously. These requirements have been a motivation for innovation in local loop. There are two remarkable challenges in local loop technologies.

One is mainly from the expansion of landscape in service types. According to drastic growth of Internet, to access Internet at home (or office) became a usual lifestyle today. Moreover, to enjoy multimedia services at home will not be strange in near future. These services require broadband local loop systems. To deal with this situation in short term, the *digital subscriber line* (DSL) technologies, including *high-bit-rate DSL* (HDSL), *asymmetrical DSL* (ADSL), and *very-high-bit-rate DSL* (VDSL), have been studied and developed [1].

Another technical advance, on which we will focus in this chapter, is the *wireless local loop* (WLL) adopting radio as the transmission medium. WLL is often called the *radio local loop* (RLL) or the *fixed wireless access* (FWA). And WLL services are also referred to as the 'fixed cellular services'. WLL has many advantages from the viewpoints of the service providers and subscribers [2–8]:

- The WLL approach significantly speeds the installation process since it can eliminate the wires, poles, and ducts essential to wired networks. Thus, WLL systems can be rapidly developed, easily extended, and are distance insensitive. Since WLL is a quick start for startup systems, wireless access is viable means to meet the high demand for POTS in many developing countries.

- The operations and maintenance are easy and the average maintenance time per subscriber per year is short (40 min compared to 2.2 hr for wireline).
- Using advanced digital radio technologies, WLL can provide a variety of data services and multimedia services as well as voice.
- Among radio systems, WLL enjoys the merits of fixed system: using high-gain directional antennas, the interference decreases. This reduces the frequency re-use distance, increases the possible number of sectors in a sectored cell, and increases, in turn, the system capacity.

Since WLL is a kind of radio system, it is natural that the WLL technologies have been affected by wireless mobile communication technologies. In fact, as will be shown later, most of WLL systems are developed according to the standards (or their variants) for mobile systems. In terms of multiple access technique, they have adopted the *frequency-division multiple access* (FDMA), the *time-division multiple access* (TDMA), the *code-division multiple access* (CDMA), or their hybrids.

Basically, almost all of wireless systems or multiple access techniques can be used for WLL. However, it is also true that there exist some technologies or systems that have comparative advantages in a certain WLL environment. The reference frame for comparison between systems is given by the service requirements in a specific service area. In this chapter, we investigate WLL services. The insight into WLL services gives:

- as mentioned above, the reference for comparison between several systems or technologies available for WLL, and
- some prospects of future WLL services that can be a motivation for further study and development of systems.

This chapter is organized as follows. Section 13.2 gives an overview of a typical architecture of WLL systems. Section 13.3 explains the WLL service requirements in developing and developed countries. Section 13.4 outlines the services provided by the representative systems available today or soon to be available. Section 13.5 discusses some further considerations for future WLL services with the relevant technologies. Finally, the chapter is concluded with Section 13.6.

13.2 WLL System Architecture

Since WLL systems are fixed, the requirement for interoperability of a subscriber unit with different base stations is less stringent than that for mobile services. As a result, there exist a variety of standards and commercial systems. Each standard (or commercial system) has its own air-interface specification, system architecture, network elements, and terminology. Moreover, although the network elements in different systems have same terminology, the functions of the elements may differ according to systems. In this section, we present a conceptual (and typical) architecture of WLL systems (see Figure 13.1).

The *fixed subscriber unit* (FSU) is an interface between subscriber's wired devices and WLL network. The wired devices can be computers or facsimiles as well as telephones. Several systems use other acronyms for FSU such as the *wireless access fixed unit* (WAFU), the *radio subscriber unit* (RSU), or the *fixed wireless network interface unit* (FWNIU).

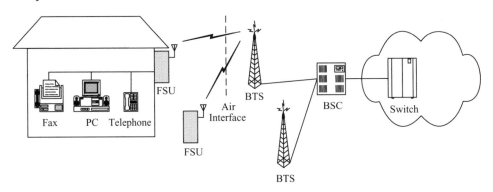

Figure 13.1 Typical architecture of WLL

FSU performs channel coding/decoding, modulation/demodulation, and transmission/reception of signal via radio, according to the air-interface specification. If necessary, FSU also performs the source coding/decoding. We will discuss some air-interface specifications later, but detailed description of them is beyond the scope of this chapter.

When a dummy telephone set is used, FSU may perform dial-tone generation function for users so as not to be aware of WLL system. FSU also supports the computerized devices to be connected to the network by using voice-band modems or dedicated data channels.

There are a variety of FSU implementations. In some types of commercial products, an FSU is integrated with handset. The basic functions of this integrated FSU are very similar to those of handset for mobile communications, except that it does not have a rich set of functions for mobility management. Another example of FSU implementation is a high-capacity, centralized FSU serving more than one subscriber. Typical application of this type of FSU can be found in business buildings, apartment blocks, and the service area where some premises are located near by (see Figure 13.2).

FSU is connected with the base station via radio of which band is several hundreds of MHz or around 2 GHz. Since WLL is a fixed service, high-gain directional antennas can be used between FSU and the base station, being arranged by line-of-sight (at least, nearly). Thus, WLL signal channel is a Gaussian noise channel or strong Rician channel (not a Rayleigh fading channel) [7]. This heightens drastically the channel efficiency and the capacity of the system.

The base station is implemented usually by two parts, the *base station transceiver system* (BTS) and the *base station controller* (BSC). In many systems, BTS performs channel coding/decoding and modulation/demodulation as well as transmission/reception of signal via radio. BTS is also referred to as the *radio port* (RP) or the *radio transceiver unit* (RTU).

A BSC controls one or more BTSs and provides an interface to the local exchange (switch) in the central office. An important role of BSC is to transcode between the source codes used in wired network and that at the air-interface. From the above roles, a BSC is often called the *radio port control unit* (RPCU) or the *transcoding and network interface unit* (TNU).

WLL systems do not need to offer mobile services basically, even if some systems provide limited mobile services. Thus, for example, there is no *home and visitor location register* (HLR/VLR) in a WLL system and its overall architecture may be simpler than that of the mobile systems.

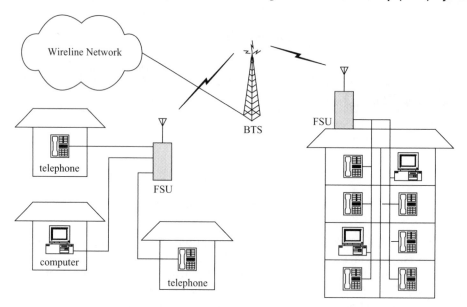

Figure 13.2   Fixed subscriber unit serving multiple subscribers

As one can easily guess from Figure 13.1, the WLL services depend not only on the functionality of FSU, BTS, BSC, and air-interface specification but also on the service features provided by the switch in the central office. For example, when WLL is used as a telephony system, there are the basic telephony services (e.g. call origination, call delivery, call clearing, emergency call, etc.) and the supplementary services (e.g. call waiting, call forwarding, three-way calling, calling number/name delivery, etc.). In addition, as in the wired systems, the features such as custom calling features, Centrex features, *custom local area signalling services* (CLASS) can be supported by the switch [5,7]. If the air-interface provides a transparent channel to the switch, these service features depend totally on the switch functions. So, we hereafter focus on the air-interface specifications related to WLL services rather than the service features by the switches.

13.3   WLL Service Requirements

The communication service requirements depend heavily on the socio-economical situations of the service areas. In general, the WLL services required in developing countries and/or regions can differ from that in developed ones.

13.3.1   Developing Countries/Regions

In many developing countries and/or regions, the infrastructures for basic telephone services are still insufficient. Accordingly, a lot of population in these areas has not been served with even POTS. This reduces the opportunity for the people to acquire information and, further, deepens the problem of information inequality.

For these areas, the emphasis points of WLL service requirements can be summarized as follows:

- In terms of service coverage, a wide area should be covered within relatively short period.
- Especially, for the regions with dense population, a high-capacity system is indispensable. Here, the capacity means the available number of voice channels for given bandwidth.
- On the other hand, there may exist wide areas with sparse population. For these service areas, if a small population with low traffic load resides near by, a centralized FSU serving more than one subscriber can be a solution (see Figure 13.2).
- The service fee per subscriber must be low so as to offer the universal service. For this, a high-capacity system is again needed and the cost of system implementation and operation should be low.
- The system should be implemented rapidly so that the services might be launched quickly. In choosing systems, the possibility of the rack of social overhead capitals (e.g. loads or electronic power) in some areas also should be considered [9].

As a trade-off to fulfil the requirements of high-capacity with low service fee, a medium-quality and relatively low data rate of channel (typically, up to 16 kbps) may be unavoidable. Using this channel, only voice and/or voice-band low-speed data communications are possible. However, the service requirements to the advanced services (e.g. high-speed data and broadband communications) will arise after (or with) the penetration of POTS. Therefore, at the initial choice and installation of WLL system, the service provider should take into account the future evolution of system to provide advanced services.

13.3.2   Developed Countries/Regions

In the developed countries and/or regions, the service requirements contain not only POTS but also other advanced services. It is usual that more than one local switching service providers and cellular mobile service providers coexist in these service areas. We examine the WLL service requirements from the standpoint of each service provider.

WLL provides a means to establish local loop systems, without laying cables under the ground crowded with streets and buildings. Thus, WLL is regarded as one of the most attractive approaches to the second local switching service providers. Unfortunately from the second providers' perspective, there are one or more existing providers (i.e. the first providers) who have already installed and operated wireline networks. To meet the increasing and expanding users' service requirements for high-speed data and multimedia services as well as voice, the first providers try to evolve their networks continually (for example, using DSL technologies). The second providers, entering the market in this situation, should offer the services containing competitive ones in terms of service quality, data rate of channel, and supplementary services, etc. That is, the WLL channel of the second provider should be superior to or, at least, comparable with the first operators' one in quality and data rate. Therefore, WLL should provide toll quality voice and at least medium-speed data corresponding to the *integrated services digital network* (ISDN) *basic rate interface* (BRI, 2B + D at 144 kbps). In addition, to give subscribers a motivation to

Table 13.1 Comparison of public wireless communications services [4]

	WLL Only Service	Cellular Mobile Service (including High-Tier PCS)	Mobile/WLL Bundled Service
Network Elements	Local Exchange/BSC/BTS FSU	MSC/BSC/BTS Mobile handset	MSC/BSC/BTS Mobile handset and FSU
Subscriber Unit	FSU with wireline feeling (e.g. dial-tone)	Mobile handset	Mobile handset, FSU with wireline feeling
Services	No (or low) mobility, Medium- to high-speed data, Wireline-like supplementary services	High mobility, Low- to medium-speed data	Low/High mobility, Low- to medium-speed data, Wireline-like supplementary services

migrate to the new provider, the service fee of the second provider needs to be lower than that of the first operators.

Even to the first local switching service providers having wireline networks, WLL can be a useful alternative for their network expansion. Most countries impose the *universal service obligation* (USO) upon the first operators. In this case, WLL can be considered as a supplementary means to wireline networks, for covering areas with sparse population, e.g. islands. The first service requirement for this application of WLL is the compatibility with and the transparency to the existing wireline network.

On the other hand, the cellular mobile service providers can offer easily WLL services by using their existing infrastructure for mobile services. In this case, *fixed* WLL service may have competitiveness by combining with the *mobile* services. For example, these two services can be offered as a bundled service [4,7]. In addition, so-called 'one-phone service' can be offered with an appropriate billing strategy. That is, with a single subscriber unit, a subscriber enjoys the fixed WLL services at home and the mobile services on the street.

Table 13.1 gives a comparison between the WLL services using a dedicated network and the mobile/WLL bundled services. Note that the table also contains the pure cellular mobile services for the purpose of comparison.

13.4 Services with Currently Available Systems

Now let us examine the WLL services provided with the systems that follow international (or domestic) standards and are commercially available today. We focus on the standards rather than specific products.

Most of WLL systems are developed according to the standards (or their variants) for mobile systems. Among mobile systems, we in this chapter consider only the digital systems (the second- and the third-generation systems) although the analogue systems, e.g. the *advanced mobile phone service* (AMPS) systems, still are used in many regions. The mobile systems can be categorized into the low-tier and the high-tier systems.

Low-tier systems support only low mobility in general. They can be characterized by low transmission power, small coverage with a base station, high subscriber density, and high circuit quality services. In comparison with high-tier systems, low-tier systems provide more wireline-like services. There are several standards for low-tier systems. The representative examples are the *personal access communications system* (PACS), the *digital enhanced cordless telecommunications* (DECT), and the *personal handy-phone services* (PHS). These standards adopt the TDMA technology.

The high-tier cellular systems support high mobility and can be characterized by the wider coverage with relatively low data rate. These systems include the second-generation digital cellular systems using 800 MHz band (e.g. IS-95 CDMA, IS-54 TDMA, and GSM) and their up-banded variations for the *personal communications services* (PCS) using $1.8 \sim 2.0$ GHz band (for example, ANSI J-STD-008 CDMA as an up-banded version of IS-95).

Among the above-mentioned systems, we briefly outline PACS, DECT, and IS-95 CDMA systems. We also review a wide-band CDMA system recently commercialized.

13.4.1   PACS Services [8,10]

PACS [11] is a hybrid standard of Bellcore's *wireless access communications system* (WACS) and Japanese PHS.

In PACS, the basic unit of channel is a time slot per TDMA frame, which transports data at 32 kbps. PACS offers toll-quality voice using ITU-T G.726, *adaptive differential pulse code modulation* (ADPCM), at 32 kbps as the default coding scheme. Optionally, 16 kbps *low-delay code-excited linear prediction* (LD-CLEP) being defined as ITU-T G.728 can be used.

For voice-band data, PACS provides 64 kbps *pulse code modulation* (PCM) connection (ITU-T G.711) by aggregating two time slots. This service is used to support all voice-band modems including 56 kbps modems.

PACS supports circuit mode and packet mode data services. In addition, individual message service and interleaved speech/data service are also provided.

- *Circuit-mode data service*: PACS offers reliable real-time data transport service using *link access procedure for radio* (LAPR). LAPR operating in a 32 kbps channel provides a data throughput of more than 28 kbps at wireline error rate (10^{-6}).
- *Messaging services*: This is two-way point-to-point message service for large file transfer up to 16 Mbytes. The messages can contain text, image, audio, and video files.
- *Packet-mode data service*: This is a shared, contention-based, RF packet protocol using a data sense multiple access contention mechanism. It supports FSU by using single time slot (32 kbps) or multiple time slots (up to 256 kbps) per TDMA frame. The applications being suitable over the PACS packet channel are wireless Internet access and mobile computing, etc.
- *Interleaved Speech/Data*: It provides the ability to transmit both speech information and data information by using a single 32 kbps time slot. Data are transmitted during the silent period of voice.

13.4.2   DECT Services

DECT standard [12] has been developed to replace the *cordless telephone-2* (CT-2) standard. It originally supports small cells (radius of $100 \sim 150$ m) with pedestrian-speed mobility. To use DECT in WLL applications, one of the most important problems to be solved is to extend the maximum coverage of a fixed part (i.e. BTS). A solution is to use directional antennas, by which the maximum diameter of a cell can be extended up to several kilometers. For rural applications, the coverage can be extended by using repeaters at the expense of capacity [10].

The basic unit of channel in DECT is a time slot per TDMA frame, operating at 32 kbps. If data rates higher than 32 kbps are required, multiple time slots per frame are used. Otherwise, if the requested data rate is lower than 32 kbps, several FSUs can share a 32 kbps channel by skipping time slots. DECT offers toll quality digital speech and voice-band modem transparency either via a 32 kbps ADPCM codec (ITU-T G.726) or as a 64 kbps PCM (ITU-T G.711) bearer service [13]. As of 1997, DECT provides 480 kbps full duplex data transfer and basic ISDN access [1]. Since all user information is encrypted, there is confidentiality between the different users belonging to a same cell. DECT has signalling compatibility with basic ISDN and GSM. For more detailed aspects of DECT WLL, one can refer to [14].

13.4.3 IS-95 CDMA

IS-95-A [15] standard has been developed for a digital cellular system with *direct sequence* (DS) CDMA technology, operating at 800 MHz band. ANSI J-STD-008 [16] being an up-banded variation of IS-95 is a standard for PCS systems, operating at 1.8 ~ 2.0 GHz band. Recently, IS-95-B [17] merges IS-95-A and ANSI J-STD-008.

In DS-CDMA systems, each channel is identified by its unique spreading code. IS-95 based CDMA WLL can support two rate sets. A code channel (that is, a traffic channel) operates at maximum of 9.6 kbps with the rate set 1 or 14.4 kbps with rate set 2. Using rate set 1 (rate set 2), the system supports 8 kbps (13 kbps) *Qualcomm's codebook excited linear predictive* (QCELP) vocoder.

This channel can be used for circuit-mode data transmission. Since a type of *radio link protocol* (RLP) [18] is used in this case, the effective data rate lowers at around 7.2 kbps for 9.6 kbps channel. The packet-mode data transmission is also possible, although it is not popular still nowadays.

IS-95-B offers high-speed data services through code aggregation. In IS-95-B systems, multiple codes (up to eight codes) may be assigned to a connection. Thus, the data rate is maximum of 76.8 kbps using rate set 1 or 115.2 kbps using rate set 2. Since IS-95-B can be implemented without changing the physical layer of IS-95-A [19], it is relatively easy for the vendor of IS-95 WLL system to develop the IS-95-B WLL system.

In mobile IS-95 systems, a sectored cell is designed with three sectors in usual. As mentioned above, in WLL systems, the antennas for BTS and FSU can be arranged by line-of-sight and this reduces interference from the other user. So, the CDMA WLL cell can be designed with six sectors [7]. This increases the frequency efficiency and the system capacity.

13.4.4 Service using Wide-band CDMA Systems

The existing cellular systems (including the second-generation digital systems) have some limitations in supporting high-speed data or multimedia services because of its insufficient maximum data rate per channel. An alternative technology to cope with this problem is the *wide-band CDMA* (W-CDMA). In comparison with the existing narrowband CDMA systems (e.g. IS-95 system with the spreading bandwidth of 1.25 MHz), W-CDMA systems use higher chip rate for direct sequence spread spectrum and, thus, spread its information into wider spectrum bandwidth (typically, equal to or over 5 MHz). Thus, data rate per code channel in W-CDMA can be higher than that in narrowband system. Note that all of the major candidates for *radio transmission technology* (RTT) of the international mobile telecommunications-2000 (IMT-2000) systems have proposed W-CDMA for next-generation mobile communication systems (e.g. [20–22]). Therefore, to try to use W-CDMA systems for WLL application is natural to the second operators, entering the market newly. In this section, we explain the W-CDMA WLL services in Korea as an example.

In Korea, the development and the standardization of W-CDMA systems both for mobile service and for WLL service are conducted simultaneously. As a result, several vendors have developed W-CDMA systems according to Korean WLL standard [23] and a second local switching service provider in Korea has a plan to start WLL service with these systems in 2000. WLL services with these systems are as follows.

The downlink (from BTS to FSU) uses the band from 2.30 to 2.33 GHz and the uplink (from FSU to BTS) uses the band 2.37 ~ 2.40 GHz. Thus, the bandwidth of each link is 30 MHz. The spreading bandwidth can be either 5 MHz or 10 MHz. For both spreading bandwidth, the information bit rates are 8, 16, 32, 64, and 80 kbps. For the case of 10 MHz spreading bandwidth, 144 kbps of information bit rate is also available.

The WLL standard defines several options for voice codec: 64 kbps PCM (ITU-T G.711), 32 kbps ADPCM (ITU-T G.726), 16 kbps LD-CELP (ITU-T G.728), and 8 kbps conjugate structure algebraic-code-excited linear prediction (CS-ACELP, ITU-T G.729). However, the service provider seems to offer voice services using 16 kbps LD-CELP and 32 kbps ADPCM since those give toll quality of voice with adequate system capacity. As the voice-band data services, G3 facsimile and 56 kbps modem are planned.

For packet mode data transmission, some dedicated channels, which are separated from voice channels, are provided. They are the packet access channels in uplink and the packet traffic channels in downlink. Using these channels, packet data services up to 128 kbps are offered. In addition, ISDN BRI is also provided.

13.4.5 Comparison of Services

Table 13.2 summarizes the WLL services mentioned above. As shown in the table, most systems offer toll quality voice services and medium- to high-rate data services. The applications being suitable over this data channel are wireless Internet access, mobile computing, and file retrieval services.

Among these systems, PACS and DECT, based on low-tier cellular system technologies, seem to be suitable for urban and developed region since they support relatively high quality channel in the small ranges.

On the other hand, the IS-95 CDMA systems being currently used (i.e. based on TIA/EIA/IS-95-A [15] or ANSI J-STD-008 [16]) have higher capacity than TDMA systems [7] and support wider service range per BTS. Thus, CDMA seems to be more appropriate choice for rural area and for developing regions. In the developed countries, the IS-95 cellular mobile service providers can offer WLL services also, using same infrastructure.

Table 13.2   Summary of WLL services

	PACS	DECT	IS-95	W-CDMA
Voice Codec	32 kbps ADPCM (16 kbps LDCELP)	32 kbps ADPCM 64 kbps PCM	13 kbps QCELP (8 kbps QCELP)	32 kbps ADPCM 16 kbps LD-CELP
Voice-Band Data	voice-band modem up to 56 kbps	voice-band modem transparency via voice channel	14.4 kbps (9.6 kbps)	voice-band modem up to 56 kbps
Data Service: Rate per Connection	32 ~ 256 kbps	up to 480 kbps* Basic ISDN access	up to 115.2 kbps** (76.8 kbps)	up to 128 kbps (144 kbps BRI)
Coverage per BTS	Small	Small	Large	Medium to Large
Capacity per Cell	Medium	Medium	High	High

* as of 1997.
** when IS-95-B is used.
( ) denotes the rate set 1 for IS-95, or the optional items in others.

In this case, the fixed WLL service and mobile service can be a bundled service in urban area as mentioned before. As another strategy, the provider may offer the mobile services in urban area and the WLL services in rural area.

IS-95-B and W-CDMA systems taking advantages of state-of-the-art technologies can be used in any region, because of their high-capacity, wide service range per BTS, and high channel quality. A demerit of these systems is that the technology is not proven sufficiently still in commercial experiments.

However, all of the systems discussed in this section cannot offer the local loops for future multimedia services such as video. One of the reasons is that these systems are originally based on the mobile systems technologies. We will review another alternative to satisfy these requirements in the next section.

13.5   Further Considerations for Multimedia Services

For bandwidth hungry services, such as video-telephony or video-on-demand (VOD), the systems mentioned in the previous section are not sufficient. The constraint on capacity per channel can be relieved by migrating to higher frequency ranges and applying broadband wireless systems [5]. In fact, WLL concept in wide-sense contains not only the systems mentioned in previous section but also the microwave *multipoint distribution services* (MMDS), the local *multipoint distribution services* (LMDS), the *wireless asynchronous transfer mode* (WATM), and the satellite access.

In many countries, LMDS, which is often called the *local multipoint communication system* (LMCS) [24], is considered as a strong candidate for next-generation *broadband WLL* (B-WLL) services. The spectrum for LMDS differs from country to country but it is usually 20 ~ 30 GHz band. The main purpose of LMDS in its beginning period is the wireless distribution of television program. Now, the LMDS applications include a variety of multimedia services such as POTS, ISDN, *broadband ISDN* (B-ISDN), television program distribution, video conference, VOD, teleshopping, and Internet access. In Korea, for instance, a standardization process for B-WLL using LMDS technology is in progress since 1997. A candidate standard [25] is based on DAVIC specification [26] adopting LMDS as the local access technology.

One of most serious problems in wireless multimedia services is traffic asymmetry between uplink and downlink [20,21,27,28]. For example, let us consider Internet access or remote computing. In these applications, short commands are transmitted via uplink, whereas relatively large files are transmitted via downlink. In these cases, if both links have the same bandwidth, the system capacity can be limited by the downlink. This, in turn, results in bandwidth waste of uplink and, eventually, spectrum inefficiency. To cope with traffic unbalance, the spectrum allocation for LMDS is given to be asymmetric between uplink and downlink. For example, in Korea, the uplink and the downlink bandwidths are 500 MHz (between $24.25 \sim 24.75$ GHz) and 1200 MHz (between $25.50 \sim 26.70$ GHz), respectively.

In fact, the problem caused by traffic asymmetry between uplink and downlink is inherent in any *frequency division duplex* (FDD) system. Note that LMDS is also an FDD system. To overcome asymmetry problem:

- the downlink bandwidth should be wider than the uplink one, and
- the network operator should be able to reset easily the bandwidth difference since the ratio of traffic volumes on two links depends on the constitution of traffic classes and it varies from area to area and from time to time.

An approach to do this is to use *time division duplex* (TDD) between two links. Thus, CDMA/TDD systems, having both the merits of CDMA (e.g. in capacity) and the advantages of TDD (e.g. flexibility in resource allocation), have been attracting many researchers' attention recently [20,21,28].

References

[1] M. Gagnaire, 'An Overview of Broad-Band Access Technologies,' *Proc. IEEE*, vol. 85, no. 12, pp. 1958–1972, Dec. 1997.

[2] V. K. Garg and E. L. Sneed, 'Digital Wireless Local Loop System,' *IEEE Commun. Mag.*, vol. 34, no. 10, pp. 112–115, Oct. 1996.

[3] B. Khasnabish, 'Broadband to the Home (BTTH): Architectures, Access Methods, and the Appetite for it,' *IEEE Network*, vol. 11, no. 1, pp. 58–69, Jan./Feb. 1997.

[4] H. Huh, S. C. Han and D. G. Jeong, 'WLL Services using Cellular Mobile Network,' Shinsegi Telecomm R&D/TR P2-97-04-02, Apr. 1997.

[5] H. Salgado, 'Spectrum Allocation for Fixed Wireless Access Technologies in the Americas,' in *Proc. IEEE VTC '98*, Ottawa, Canada, pp. 282–187, May 1998.

[6] A. R. Noerpel and Y. -B. Lin, 'Wireless Local Loop: Architecture, Technologies and Services,' *IEEE Personal Commun.*, vol. 5, no. 3, pp. 74–80, June 1998.

[7] W. C. Y. Lee, 'Spectrum and Technology of a Wireless Local Loop System,' *IEEE Personal Commun.*, vol. 5, no. 1, pp. 49–54, Feb. 1998.

[8] C. R. Baugh, E. Laborde, V. Pandey and V. Varma, 'Personal Access Communications System: Fixed Wireless Local Loop and Mobile Configuration and Services,' *Proc. IEEE*, vol. 86, no. 7, pp. 1498–1506, July 1998.

[9] M. P. Lotter and P. van Rooyen, 'CDMA and DECT: Alternative Wireless Local Loop Technologies for Developing Countries,' in *Proc. IEEE PIMRC '97*, Helsinki, Finland, pp. 169–173, Sep. 1997.

[10] C. C. Yu, D. Morton, J. E. Wilkes and M. Ulema, 'Low-tier Wireless Local Loop Radio Systems—Part 1: Introduction,' *IEEE Commun. Mag.*, vol. 35, no. 3, 84–92, Mar. 1997.

[11] ANSI J-STD-014, Personal Access Communications Systems Air Interface Standard, 1995.

[12] ETS 300 175, Digital Enhanced Cordless Telecommunications (DECT); Common Interface (CI), European Telecommunications Standards Institute (ETSI), 1992.

[13] D. Akerberg, F. Brouwer, P. H. G. van de Berg and J. Jager, 'DECT Technology for Radio in the Local Loop,' in *Proc. IEEE VTC '94*, Stockholm, Sweden, pp. 1069–1073, June 1994.

[14] M. Zanichelli, 'Cordless in the Local Loop,' in *Cordless Telecommunications in Europe*, W. Tuttlebee, Ed., Springer-Verlag, London, 1997.

[15] TIA/EIA/IS-95-A, Mobile Station-Base Station Compatibility Standard for Dual-Mode Wideband Spread Spectrum Cellular System, 1995.

[16] ANSI J-STD-008, Personal Station-Base Station Compatibility Requirements for 1.8 to 2.0 GHz Code Division Multiple Access (CDMA) Personal Communications Systems, 1996.

[17] TIA/EIA/SP-3693 (to be published as TIA/EIA-95), Mobile Station-Base Station Compatibility Standard for Dual-Mode Wideband Spread Spectrum Cellular System, Baseline Version, July 1997.

[18] TIA/EIA/IS-707-A.2 (PN-4145.2), Data Service Options for Wideband Spread Spectrum Systems: Radio Link Protocol, V&V version, July 1998. See also, TIA/EIA/IS-707.8 (PN-4145.8), Data Service Options for Spread Spectrum Systems: Radio Link Protocol Type 2, published version, Feb. 1998.

[19] D. N. Knisely, S. Kumar, S. Laha and S. Nanda, 'Evolution of Wireless Data Services: IS-95 to cdma2000,' *IEEE Commun. Mag.*, vol. 36, no. 10, pp. 140–149, Oct. 1998.

[20] ETSI/SMG/SMG2, The ETSI UMTS Terrestrial Radio Access (UTRA) ITU-R RTT Candidate Submission, Jan. 1998.

[21] Ad-hoc T, IMT-2000 Study Committee of ARIB, Japan's Proposal for Candidate Radio Transmission Technology on IMT-2000: W-CDMA, June 1998.

[22] TIA TR-45.5, The cdma2000 ITU-R RTT candidate submission, June 1998.

[23] TTA.KO-06.0015 and TTA.KO-06.0016, Wideband CDMA Air Interface Compatibility Interim Standard for 2.3 GHz Band WLL System, (in Korean), Telecommunications Technology Associations (TTA), Dec. 1997.

[24] G. M. Stamatelos and D. D. Falconer, 'Milimeter Radio Access to Multimedia Services via LMDS,' in *Proc. IEEE GLOBECOM '96*, London, UK, pp. 1603–1607, Nov. 1996.

[25] ETRI, Proposal for Broadband WLL (B-WLL) Standard Candidate, (in Korean), Version 1.0, Nov. 1998.

[26] DAVIC 1.2 Specification Part 8, Lower Layer Protocols And Physical Interfaces (Technical Specification) Revision 4.2, Digital Audio-Visual Council (DAVIC), 1997.

[27] J. Zhuang and M. E. Rollins, 'Forward Link Capacity for Integrated Voice/Data Traffic in CDMA Wireless Local Loops,' in *Proc. IEEE VTC '98*, Ottawa, Canada, pp. 1578–1582, May 1998.

[28] D. G. Jeong and W. S. Jeon 'CDMA/TDD System for Wireless Multimedia Services with Traffic Unbalance between Uplink and Downlink,' *IEEE J. Select. Areas Commun.*, vol. 17, no. 5, pp. 939–946, May 1999.

Index